PORSCHE **MUSEUM**

Ferdinand Porsche 1950 am Großglockner mit einem »VW-Käfer« aus der Serienproduktion
Ferdinand Porsche at the Grossglockner in 1950, with a series-production B

PROLOG FERDINAND PORSCHE UND DER VOLKSWAGEN

FERDINAND PORSCHE AND THE VOLKSWAGEN

Der Name Porsche ist bereits seit Beginn des letzten Jahrhunderts mit wegweisenden Innovationen im Automobilbau verbunden. 1900 entwickelte Ferdinand Porsche in Wien bei der k.u.k. Hofkutschenfabrik Lohner Elektromobile und Fahrzeuge mit benzin-elektrischem Antrieb – ganz nebenbei die ersten Hybrid-Automobile der Welt. Wenig später führte er den Allradantrieb und die Vierradbremse in den Automobilbau ein. 1906 wurde Ferdinand Porsche Technischer Direktor bei Austro-Daimler in Wiener Neustadt und war, gerade 31 Jahre jung, für die Modellpalette eines der führenden europäischen Fahrzeugherstellers verantwortlich. Mit dem »Sascha« kreierte er einen Kleinwagen, der 1922 bei der Targa Florio als Klassensieger in die Motorsportgeschichte einging. Nur ein Jahr später wechselte der inzwischen weit über

Since the beginning of the previous century, the Porsche name has been associated with pioneering innovations in automobile design and construction. In 1900 Ferdinand Porsche was employed by the k.u.k. Hofkutschenfabrik Lohner, suppliers to the Emperor and royalty in Vienna, and not only developed electric vehicles but also the concept of combining a gasoline engine and an electric motor – which happens to make him the inventor of the world's first hybrid! Only a short time later, Porsche was designing vehicles with all-wheel drive and – a complete novelty at the time – four-wheel brakes. In 1906 he moved to Austro-Daimler in Wiener Neustadt as Technical Director, and at the remarkably early age of 31 was responsible there for the model program of a leading European carmaker. One of his designs, the "Sascha"

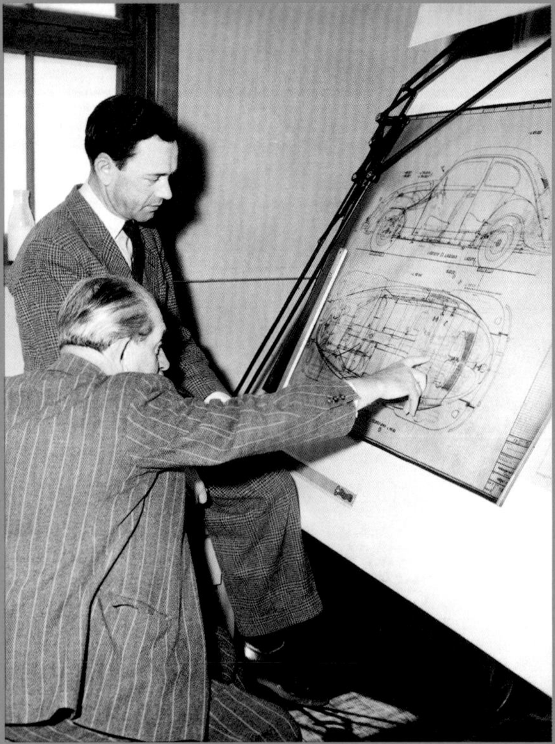

Ferdinand Porsche (sitzend) und Sohn Ferry, 1950
Ferdinand Porsche (sitting) and his son Ferry, 1950

die Automobilszene hinaus bekannte Konstrukteur nach Stuttgart zur Daimler-Motoren-Gesellschaft, wo es vor allem die kompressoraufgeladenen Sport- und Rennwagen waren, die seinen Ruf als bahnbrechenden Automobilbauer weiter in die Welt hinaus trugen. Als Krönung seiner Laufbahn eröffnete er am 25. April 1931 in Stuttgart sein eigenes Konstruktionsbüro. Das als »Dr. Ing. h.c. F. Porsche GmbH, Konstruktion und Beratung für Motoren- und Fahrzeugbau« firmierende Unternehmen wurde zu einer Keimzelle der Automobiltechnik. Das kleine Ingenieurs-Team um Ferdinand Porsche entwickelte Fahrzeuge und technische Detaillösungen für nahezu alle großen deutschen Hersteller. Schon in den Anfangsjahren war das Spektrum der Auftragskonstruktionen weit gefächert und reichte von der Wanderer-Limousine Porsche-Typ 7 bis zum Typ 22, einem Grand-Prix-Rennwagen für die Auto Union.

Die Entwicklung des Volkswagens, intern Typ 60 genannt, war ein zentrales Ereignis in der Frühgeschichte des Unternehmens. Im Januar 1934 verfasste Ferdinand Porsche ein »Exposé betreffend den Bau eines Deutschen Volkswagens«, das die Automobilgeschichte nachhaltig verändern sollte. Das preiswerte und

small car, has gained a place in automobile history by winning its class in the 1922 Targa Florio road race. A year after this, Ferdinand Porsche moved to the Daimler-Motoren-Gesellschaft in Stuttgart. By now his reputation was firmly established beyond the automobile industry as well. For Daimler, he was mainly responsible for the supercharged sports and racing cars, which helped to make his name as an innovative car designer even better known throughout the world.

He took an ultimate career step on April 25, 1931 when he opened his own design and consulting engineer's office in Stuttgart, "Dr. Ing. h.c. F. Porsche GmbH, Konstruktion and Beratung für Motoren- and Fahrzeugbau". This company proved to be a breeding ground for advanced automobile technology. The small team of engineers working with Ferdinand Porsche developed complete vehicles and technical details for almost all the major German carmakers. Even in the company's early years it undertook a big range of design contracts, from the Wanderer sedan (the Porsche Type 7) to the Auto Union Grand Prix racing car (Porsche Type 22).

Ferdinand Porsche mit einem VW38 1940 in Wien
Ferdinand Porsche with a VW38 in Vienna, 1940

kompakte Automobil wurde zum Symbol und zu einer Ikone der Massenmotorisierung. Auch wenn einzelne technische Merkmale des späteren Volkswagens zum Teil schon vor 1934 von anderen Konstrukteuren erdacht waren, ist es das unzweifelhafte Verdienst Ferdinand Porsches, den Wagen nicht nur konstruiert, sondern ihn zum Prototyp entwickelt und gegen alle Widerstände zur Serienreife gebracht zu haben. Ohne Porsche gäbe es keinen Volkswagen. Nicht das Fahrzeug und auch nicht das Werk, dessen Planung Ferdinand Porsche ab 1937 verantwortete und das sein Schwiegersohn Anton Piëch ab 1941 als Hauptgeschäftsführer leitete.

Seine Bestimmung, ein »Wagen für das Volk« zu sein, konnte der Volkswagen erst nach dem Zweiten Weltkrieg erfüllen. Für Ferdinand Porsche war die zivile Nachkriegsfertigung seiner Konstruktion eine große Genugtuung. Während einer Autobahnfahrt mit seinem Neffen und Privatsekretär Ghislaine Kaes stellte er 1950 zufrieden fest, dass sechs von zehn entgegenkommenden Fahrzeugen Volkswagen waren. Der Volkswagen, auch liebevoll »VW-Käfer« genannt, wurde auf der ganzen Welt so populär wie kaum ein anderes

Development of the “Volkswagen” or “People’s Car”, known internally as the Porsche Type 60, was a key event in the Porsche company’s early history. In January 1934 Ferdinand Porsche wrote an “Exposé on the Production of a German People’s Car”. This document, as it transpired, was to change automobile history. The compact, low-priced car that eventually appeared was to become a symbol and an icon of mass car ownership. Other designers had thought out certain technical aspects of the later “Volkswagen” before 1934, but it was undoubtedly Ferdinand Porsche’s achievement to have drafted out the design, built the prototypes and brought the car to production maturity despite all the obstacles that had to be overcome. Without Porsche, there would be no Volkswagen: neither the car itself nor the factory in which it was to be built. Ferdinand Porsche was responsible for planning this too from 1937 on, and his son-in-law Anton Piëch became its principal General Manager in 1941.

It was only after the Second World War, however, that the Volkswagen was able to fulfill its true purpose as a “people’s car”. It gave Ferdinand Porsche great

Automodell. Und auch in punkto Produktionsdauer und Stückzahl war er ein Spitzenreiter: Erst im Juli 2003 wurde die Produktion der letzten Volkswagen, die noch in Mexiko vom Band liefen, eingestellt. Mit 21,5 Millionen Exemplaren gehört der VW-Käfer mit Abstand zu den meist gebauten Autos der Welt.

Ferdinand Porsche 1937
Ferdinand Porsche 1937

satisfaction to see his design enter production for civil purposes. Driving along a German 'autobahn' in 1950, accompanied by his nephew and private secretary Ghislaine Kaes, he was delighted to see that six out of every ten cars coming the other way were Volkswagens.

Scarcely any other car in the world was as popular as the Volkswagen, which in the course of time acquired the affectionate nickname 'Beetle', It also broke all records for its production period and production volume: the Beetle continued to leave the assembly lines in Mexico until July 2003, by which time twenty-one and a half million of these cars had been built, making the VW Beetle easily the highest-volume individual model that the car industry has ever produced.

2

PROLOG | FERDINAND PORSCHE UND DER VOLKSWAGEN

PROLOG | FERDINAND PORSCHE AND THE VOLKSWAGEN

DAS VOLKSWAGEN-EXPOSÉ
THE PEOPLE'S CAR EXPOSÉ

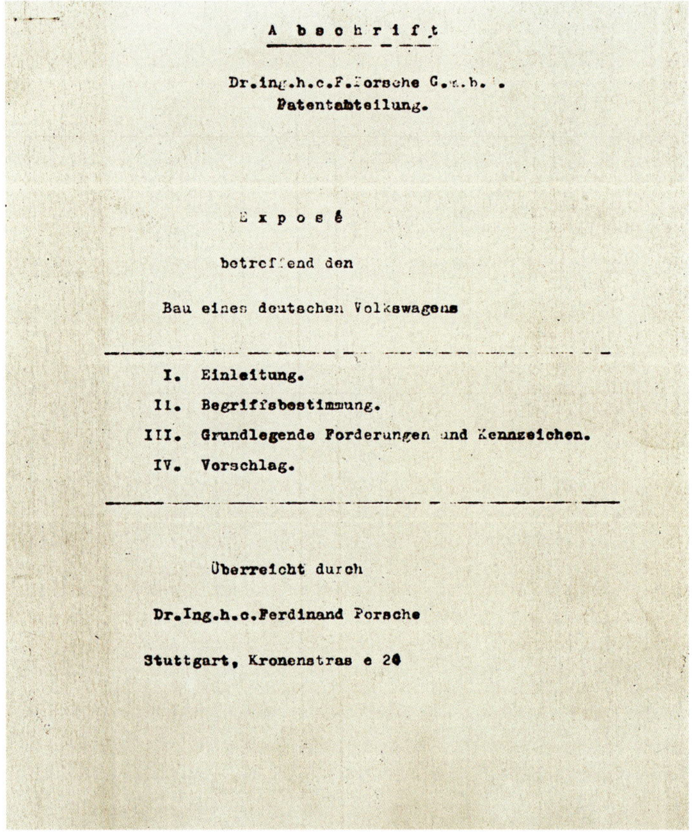

A b s c h r i f t

Dr.ing.h.c.F.Porsche G.m.b.H.
Patentabteilung.

E x p o s é

betreffend den

Bau eines deutschen Volkswagens

I. Einleitung.
II. Begriffsbestimmung.
III. Grundlegende Forderungen und Kennzeichen.
IV. Vorschlag.

Überreicht durch

Dr.Ing.h.c.Ferdinand Porsche

Stuttgart, Kronenstras e 24

DAS VOLKSWAGEN-EXPOSÉ
THE PEOPLE'S CAR EXPOSÉ

Die Idee eines »Volksautos« war in Deutschland seit der Mitte der 1920er Jahre ein Thema engagierter Diskussionen. Infolge der Wirtschaftskrise beschäftigten sich zahlreiche Ingenieure mit der Konstruktion von preisgünstigen Kleinwagen für die nach Mobilität verlangende Mittelschicht. In verschiedenen Automobilzeitschriften wurden Abhandlungen zu diesem Thema veröffentlicht, darunter richtungweisende Beiträge von Hans Ledwinka, Gustav Röhr, Béla Barényi oder Josef Ganz. Zu den wichtigsten Merkmalen dieser neuen Kleinwagenkonzepte zählten unter anderem eine Stromlinienform der Karosserie, ein Plattformrohrrahmen oder die Platzierung des Antriebsaggregates im Fahrzeugheck. Als Inhaber eines selbstständigen Konstruktionsbüros hatte auch Ferdinand Porsche diese technische Debatte aufmerksam beobachtet,

Links: Ferdinand und Ferry Porsche (1934)
Rechts: Titelseite des Volkswagen-Exposé
Left: Ferdinand and Ferry Porsche (1934)
Right: Front page of the Volkswagen exposé

The concept of a people's car had been discussed enthusiastically in Germany since the mid-1920s. Then came the world economic crisis, and many engineers felt that small, low-priced cars should be designed to satisfy the needs of middle-class buyers who were demanding improved mobility. This topic was discussed in various car magazines, with useful contributions from such leading figures as Hans Ledwinka, Gustav Röhr, Béla Barényi and Josef Ganz. Among the features which it was felt that a new type of small car should have were a streamlined body, a platform chassis and a rear engine. Now the proprietor of an independent design office, Ferdinand Porsche followed this technical debate closely; at various stages in his career he had already worked on a number of small-car designs.

Das Konstruktionsbüro
Porsche in der Stuttgarter
Kronenstraße 24
Porsche's engineering office, based in Kronenstrasse 24 in Stuttgart

On January 22, 1934 he decided to sum up his personal views on the subject in an "Exposé on the Production of a German People's Car", which he submitted to the German Reich's Transport Ministry. His design study was in fact the eighth compact car for which he had been responsible, starting with the Lohner-Porsche "Voiturette" in 1901 and continuing through the Austro-Daimler "Maja" (1906) and "Sascha" (1922) models, the Mercedes-Benz "5/25 PS" (1926/28) and the "Steyr XXX" (1930). It is true to say that the notion of a compact car ran like a thread through Ferdinand Porsche's entire previous career.

In the 1920s, when Germany was plagued by a recession, several manufacturers had tackled the task of building a moderately priced small car, for example Opel's Type 4/12 PS, known as the "Tree Frog", or the Hanomag Type 2/10 "Kommissbrot" ("Hard Tack"). But these and others failed to eliminate a prevailing problem: the low price could only be achieved by reducing the car's practical value. The Volkswagen Type 60 aimed to resolve this dilemma: it was to be a full four-seat car, cheap to build and offered at a price that most people could afford. In his exposé Ferdinand

denn innerhalb seiner beruflichen Stationen hatte er ebenfalls bereits an verschiedenen Kleinwagentypen gearbeitet. Als Ferdinand Porsche am 22. Januar 1934 schließlich seine persönliche Quintessenz zu diesem Thema als »Exposé betreffend den Bau eines Deutschen Volkswagen« dem Reichsverkehrsministerium vorstellte, war dies bereits der achte unter seiner Regie konstruierte Kompaktwagen. Angefangen bei der Lohner-Porsche »Voiturette« aus dem Jahre 1901 über die Austro-Daimler Typen »Maja« (1906) und »Sascha« (1922), den Mercedes-Benz »5/25 PS« (1926/28) und den »Steyr XXX« (1930) hatte die Kompaktwagen-Idee die gesamte bisherige Karriere Ferdinand Porsches begleitet.

In den rezessionsgeplagten Zwanziger Jahren hatten sich auch andere Hersteller mit preisgünstigen Kleinwagen beschäftigt, doch alle in diesem Zusammenhang entstandenen Wagen wie der Opel Typ 4/12 PS »Laubfrosch« oder der Hanomag Typ 2/10 »Kommissbrot« waren mit einem bis dato ungelösten Problem behaftet: Einhergehend mit dem geringen Verkaufspreis reduzierte sich stets auch der Nutzwert der Automobile. Dies sollte beim Volkswagen Typ 60 grund-

Porsche expressed his views on a future "people's car" as follows: *"In my opinion a people's car does not necessarily have to be a small car with its dimensions, power output, weight etc. artificially reduced in the misdirected manner adopted for previous efforts in this area. [...] For me, a people's car must be a full-scale, practical model that can stand comparison with any other car intended for day-to-day us. To give existing cars the status of a people's car, I believe that a fundamentally new approach has to be adopted."*

ORDER PLACED BY THE RDA

When the National Socialists seized power in Germany, mass mobility became a central objective in their propaganda. Opening the Berlin Motor Show on February 11, 1933, Adolf Hitler described for the first time his vision of "motoring for the people", taking the United States of America as his example. He pressed ahead with this plan as part of his general economic and social policy, regardless of resistance from within the automobile industry or the priorities otherwise granted to the armaments industry. Hitler called upon the German Automobile Industry Association

legend anders sein. Der VW sollte ein vollwertiges viersitziges Automobil sein, gleichzeitig aber zu einem erschwinglichen Preis produziert und angeboten werden. In seinem Exposé formulierte Ferdinand Porsche seine Ansichten über ein künftiges Volksautomobil wie folgt: *»Ich verstehe unter einem Volkswagen kein Kleinfahrzeug, das durch künstliche Verringerung seiner Abmessungen, seiner Leistung, seines Gewichtes usw. die Tradition der bisherigen Erzeugnisse auf diesem Gebiete nach der Storchenschnabelmanier weiterführt. [...] Ich verstehe unter einem Volkswagen nur ein vollwertiges Gebrauchsfahrzeug, das mit jedem anderen Gebrauchsfahrzeug gleichberechtigt in Wettbewerb treten kann. Um die bisher üblichen Gebrauchswagen zu Volkswagen zu machen, bedarf es meiner Ansicht nach grundsätzlicher Lösungen!«*

DIE AUFTRAGSERTEILUNG DURCH DEN RDA

Nach der Machtübernahme der Nationalsozialisten wurde die Massenmobilisierung ein zentrales Ziel der NS-Propaganda. Anlässlich der Eröffnung der internationalen Automobilausstellung am 11. Februar 1933 in

(the "RDA") to put his demand for a people's car into effect. On February 12, 1934 a preliminary discussion was held at the Reich's Transport Ministry, attended by representatives of various other ministries and from the RDA. At this meeting the RDA insisted that any such car should be totally practical, and that for this reason the manufacturers were opposed to the Transport Ministry's proposal to build a three-wheeled model. An important occurrence during this meeting was that Ferdinand Porsche's name was mentioned for the first time in connection with the people's car. To ensure that the project was tackled as neutrally as possible, it was to be commissioned from "one of our best independent designers". The RDA found itself in a political and strategic dilemma: on the one hand, its members wished to avoid state-subsidized competition with their own products, but on the other hand when the "Führer" expressed a wish, it had to be obeyed. To escape from this situation, the RDA declared in April 1934 that it regarded the production of a people's car as its "task by right" and that it should be carried out without funds from the government. This proviso was clearly made in an attempt to avoid state influence on the automobile industry.

Berlin beschrieb Adolf Hitler erstmals seine Vorstellung einer deutschen »Volksmotorisierung« nach amerikanischem Vorbild. Ohne Rücksicht auf Widerstände der Automobilindustrie oder rüstungspolitische Prioritäten wollte er die deutsche Massenmotorisierung als Teil seiner wirtschafts- und sozialpolitischen Pläne verwirklichen. Für die Umsetzung eines Volksautomobils nahm Hitler die deutschen Automobilhersteller in die Pflicht, die durch den »Reichsverband der Automobilindustrie« (RDA) repräsentiert wurden. Am 12. Februar 1934 fand eine erste Besprechung im Reichsverkehrsministerium zwischen verschiedenen Vertretern einzelner Ministerien und dem RDA statt. Bei diesem Treffen forderte der RDA eine uneingeschränkte Gebrauchsfähigkeit des Wagens, weshalb sich die Hersteller gegen den Vorschlag des Verkehrsministeriums aussprachen, den Volkswagen dreirädrig zu bauen. Diese Sitzung war für die Geschichte des Volkswagens von besonderer Bedeutung, da hier der Name Ferdinand Porsche zum ersten Mal genannt wurde. Um eine möglichst neutrale Behandlung der Aufgabe zu gewährleisten, sollte das Projekt »einem unserer besten, aber unabhängigen Konstrukteure« übergeben werden. Politisch wie strategisch befand

Ferdinand Porsche
im Konstruktionsbüro
Kronenstraße 24
Ferdinand Porsche in his engineering office, Kronenstrasse 24

sich der RDA allerdings in einem Dilemma: Zum einen war dem Verband nicht daran gelegen, eine staatlich geförderte Konkurrenz zu den eigenen Produkten seiner Mitgliedsfirmen heranzuziehen, andererseits konnte aber einem Wunsch des »Führers« nicht widersprochen werden. Vor diesem Hintergrund erklärte der RDA im April 1934, dass er den Bau eines Volksautomobils als seine »ureigenste Aufgabe« ansehe, die ohne öffentliche Mittel realisiert werden sollte. Letzteres vor allem, da man einen staatlichen Einfluss von der Automobilindustrie fern halten wollte.

Unter dem maßgeblichen Einfluss von Wilhelm Kissel, dem Vorstandsvorsitzenden der Daimler-Benz AG, entschloss sich der RDA schließlich dazu, den Auftrag an das neutrale Konstruktionsbüro von Ferdinand Porsche zu vergeben. Hinter Kissel stand jedoch noch ein weiterer Mann, der sich für die Beauftragung Porsches einsetzte: Jacob Werlin. In den Zwanziger Jahren war Werlin Verkaufsleiter der Benz & Cie.-Filiale in München gewesen und hatte in dieser Funktion schon früh persönliche Kontakte zum damals noch unbekannten NSDAP-Politiker Hitler geknüpft. Als überzeugter Nationalsozialist wurde Werlin zwar 1928

Wilhelm Kissel, Chairman of the Daimler-Benz AG Board of Management, exerted all his influence to persuade the RDA to commission the work from Ferdinand Porsche's independent design office. Kissel had a supporter in his efforts to have the contract awarded to Porsche: Jacob Werlin. As sales manager of the Munich branch of Benz & Cie. In the 1920s, he had established private links with Hitler at an early stage, when the NSDAP politician was still relatively unknown. A convinced National Socialist, Werlin had been dismissed by Daimler-Benz AG in 1928, but in 1933 was appointed to the Board of Management as a member without portfolio and from then on acted as the company's link with the Reich's Chancellery. Shortly after the Berlin Motor Show in March 1934, Werlin arranged a meeting between Hitler and Ferdinand Porsche. After this, the Chancellor of the German Reich was even more impressed by Porsche's technical acumen. The discussion, held at the Kaiserhof Hotel in Berlin, dealt in depth with the people's car question. Hitler was already convinced that such a vehicle would also be useful for military purposes, but had the fixed idea that the car should be powered by an air-cooled three-cylinder diesel

aus der Daimler-Benz AG entlassen, stieg jedoch 1933 zum Vorstandsmitglied ohne besonderen Geschäftsbereich auf und war von nun an die Verbindungsperson des Unternehmens zur Reichskanzlei. Kurze Zeit nach der Berliner Automobilausstellung im März 1934 arrangierte Jacob Werlin ein Treffen zwischen dem Reichskanzler und Ferdinand Porsche, das die hohe Meinung Hitlers von den technischen Fähigkeiten Porsches festigen sollte. Im Hotel Kaiserhof in Berlin wurde ausführlich über die Volkswagen-Thematik diskutiert. Hitler, der bereits an eine militärische Eignung des Fahrzeugs dachte, favorisierte als Antrieb einen luftgekühlten Dreizylinder-Dieselmotor in Sternanordnung. Auch bei der Preisstellung des künftigen Fahrzeugs hatte Hitler genaue Vorstellungen, die jedoch von allen fachkundigen Beteiligten als realitätsfern beurteilt wurden. Aus sozialpolitischen Gründen sollte der Verkaufspreis des Wagens die Obergrenze von 1.000,– Reichsmark nicht überschreiten.

Am 22. Juni 1934 schlossen der RDA und Ferdinand Porsche einen Vertrag, der die Entwicklung und den Bau eines Volkswagen-Prototypen zum Gegenstand hatte. Die Bedingungen dieser Vereinbarung waren

Chefkonstrukteur Karl Rabe (rechts) und Ferdinand Porsche
Chief Engineer Karl Rabe (right) and Ferdinand Porsche

engine with its cylinders arranged radially in a star pattern. The Führer also had a clear picture of the price to be charged for the car, although the experts attending the meeting felt this to be unrealistic: as a factor in the government's social policy, Hitler wanted the car to sell for no more than 1,000 Reichsmarks.

On June 22, 1934 the RDA concluded a contract with Ferdinand Porsche for the development and construction of a "Volkswagen" prototype. It must be said, however, that the contract's conditions were hostile to Porsche and formulated in the hope that the project would founder before long. Dr. Ing. h.c. F. Porsche GmbH was allowed a mere ten months to build a roadworthy test car. The financial reward was meager, to say the least: a fee of 200,000 Reichsmarks was to be paid to the Porsche company if the contract was completed successfully. The downpayment of only 25,000 Reichsmarks was clearly insufficient for a project of this magnitude, but it did at least rescue the Porsche company from a rather precarious financial situation. The design office at Kronenstrasse 24 in Stuttgart lost no time in allocating the internal responsibilities: Ferdinand

allerdings in der Absicht, das Projekt über kurz oder lang scheitern zu lassen, zu Ungunsten von Porsche formuliert worden. Inklusive des Baus eines fahrtüchtigen Probewagens wurde der Dr. Ing. h.c. F. Porsche GmbH eine Entwicklungsdauer von nur zehn Monaten zugestanden. Ebenso spärlich bemessen wie die Zeitvorgabe war auch die Finanzierung der Auftragsentwicklung: Insgesamt sollten 200.000,– Reichsmark als Honorar an die Firma Porsche fließen. Als Vorschuss erhielt Porsche 25.000,– Reichsmark, was zwar gemessen am Projektumfang unzureichend war, aber zunächst die Zukunft der sich in einer wirtschaftlich schwierigen Lage befindlichen Porsche GmbH sicherte. Die Arbeitsteilung innerhalb des in der Stuttgarter Kronenstraße 24 ansässigen Konstruktionsbüros erwies sich jedoch als äußerst effektiv: Während Ferdinand Porsche als Organisator und Visionär neue Ideen entwickelte und für diese Auftraggeber fand, setzten Chefkonstrukteur Karl Rabe und eine Reihe von Technikern diese im Detail um. Zu den wichtigsten Porsche-Ingenieuren zählten der Getriebe-Experte Karl Fröhlich, Josef Kales und Franz Xaver Reimspiess als Motorspezialisten, sowie Erwin Komenda, der als Stilist für die Formgebung des Volkswagens zuständig war.

Technische Zeichnung des Typ 60 vom 27.04.1934
Technical drawing of the Type 60, dated April 27, 1934

Erwin Komenda, Designer der Volkswagen-Karosserie
Erwin Komenda designed the Volkswagen car body

Porsche, the organizer and visionary, had the task of coming up with new ideas and solutions and finding clients for them; Chief Designer Karl Rabe and a team of technical experts put these ideas into effect. Among Porsche's leading engineers were the transmissions expert Karl Fröhlich, engine specialists Josef Kales and Franz Xaver Reimspiess and Erwin Komenda, the stylist responsible for the shape of the new people's car.

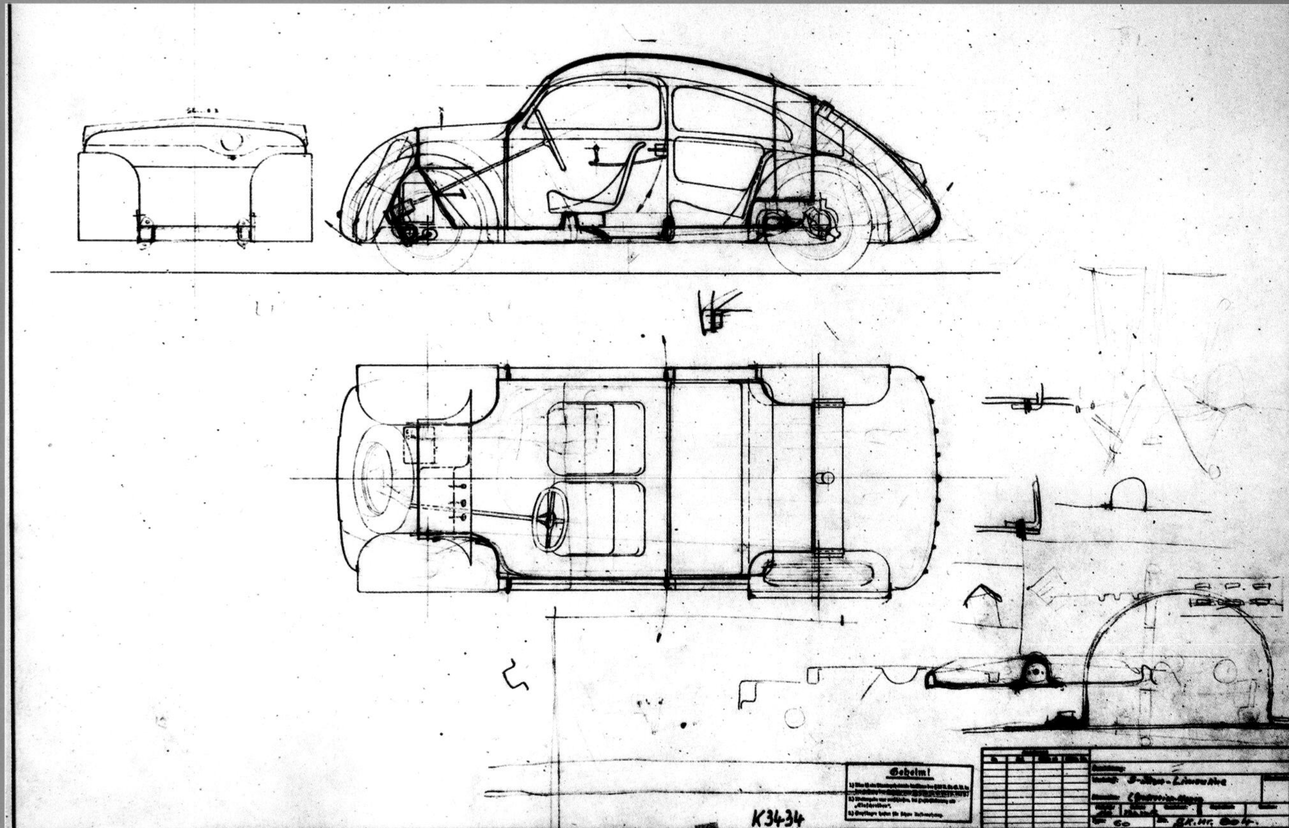
Geheim!
K3434

DIE VOLKSWAGEN-VORLÄUFER FÜR ZÜNDAPP UND NSU

PRECURSORS OF THE PEOPLE'S CAR FOR ZÜNDAPP AND NSU

IIA-20020

DIE VOLKSWAGEN-VORLÄUFER FÜR ZÜNDAPP UND NSU
PRECURSORS OF THE PEOPLE'S CAR FOR ZÜNDAPP AND NSU

Als die Porsche-Techniker im April 1934 mit der Realisierung des Volkswagen Typ 60 begannen, konnten sie bereits auf die Erfahrung von zwei ähnlichen Fahrzeugentwicklungen aus den Vorjahren zurückgreifen: die Porsche-Typen 12 und 32. Der Typ 12 war 1931/32 für den Nürnberger Motorradhersteller Zündapp entwickelt worden, der sein Verkaufsprogramm um eine eigene Automobilsparte erweitern wollte. Die im Oktober 1931 begonnene Kompaktwagen-Konstruktion besaß zwar auf Anordnung des Zündapp-Inhabers Fritz Neumeyer einen Fünfzylinder-Sternmotor, zeigte ansonsten aber eine mit dem späteren Volkswagen vergleichbare Konzeption. Neben Heckmotor und Hinterradantrieb kamen auch der Radstand von 2500 mm (VW38: 2400 mm) und die Spurweite von 1200 mm (VW38: vorn 1290 mm, hinten 1250 mm)

Porsche Typ 12 im Hof der Porsche-Villa in Stuttgart-Nord (1932)
Porsche Type 12 in the yard of the Porsche family's villa in the north of Stuttgart, 1932

In April 1934, when Porsche's technical staff began work on their Type 60, the "People's Car", they were able to base their work on experience with two similar development projects in previous years: Porsche Types 12 and 32. The Type 12 had been developed in 1931/32 for the Nuremberg-based motorcycle manufacturer Zündapp, which planned to add a carmaking division to its activities. Begun in October 1931, the new compact-car design had a five-cylinder radial engine, the brainchild of Zündapp's proprietor Fritz Neumeyer, but in other respects was similar to the later people's car concept. It had a rear engine and rear-wheel drive, and similar dimensions: the wheelbase was 2500 mm (VW38: 2400 mm) and the track 1200 mm (VW38: 1290 mm at the front and 1250 mm at the rear). The power output was 26 bhp

den Werten des Volkswagens ebenso nahe wie die Motorleistung von 26 PS (VW38: 24 PS). Die drei gebauten Zündapp-Prototypen erhielten Karosserien mit vorn angeschlagenen Türen, die von den Stuttgarter Reutter-Werken in Gemischtbauweise aus Stahl und Holz gefertigt und auf einen U-Profil-Rahmen montiert

Ferry Porsche vor der Cabriolet-Version des Zündapp-Prototyps Typ 12
The Zündapp Type 12 prototype as a convertible; Ferry Porsche is at the front

(VW38: 24 bhp). Three Zündapp prototypes were built: the bodies, built in wood and metal by the Reutter company in Stuttgart, had front-hinged doors and were mounted on a U-section chassis frame. The suspension used center-pivot swing axles and transverse leaf springs – an idea that Ferdinand Porsche had first pursued while working for Steyr.

As had been expected, road tests gave rise to numerous problems with the water-cooled radial engine; this was in unit with the gearbox at the rear of the car and had a tendency to overheat. The three-speed gearbox, with an overdrive top gear ratio, was also troublesome, with the result that in the spring of 1932 Managing Director Neumeyer decided to abandon the project. It is probable that the investment needed to start series production contributed more to this decision than the prototype's technical problems.

Another milestone on the road to the people's car was the Porsche Type 32, designed in 1933/34 for the NSU Vereinigte Fahrzeugwerke AG. In contrast to the Zündapp car, this development contract gave Ferdinand Porsche much greater scope to implement

wurden. Das Fahrwerk hatte Pendel-Schwingachsen mit Querblattfederung – eine Idee, die Ferdinand Porsche während seines Engagements bei Steyr erstmals verfolgt hatte. Während der Fahrerprobung erwies sich jedoch der wassergekühlte Sternmotor erwartungsgemäß als problematisch, da die Motor-Getriebe-Einheit im Heck eingebaut war und dort mit thermischen Problemen zu kämpfen hatte. Auch das Dreigang-Getriebe (mit Schnellgang) machte Schwierigkeiten, so dass Generaldirektor Neumeyer schließlich im Frühjahr 1932 entschied, das Fahrzeugprojekt nicht weiter zu verfolgen. Mehr noch als die technischen Probleme der Prototypen dürften jedoch die mit einem Serienanlauf verbundenen Investitionen der Grund gewesen sein, das Vorhaben aufzugeben.

Ein weiterer Meilenstein auf dem Weg zum Volkswagen war der 1933/34 für die »NSU Vereinigte Fahrzeugwerke AG« konstruierte Porsche Typ 32. Im Gegensatz zum Zündapp-Wagen hatte Ferdinand Porsche bei diesem Entwicklungsauftrag weitgehende Freiheiten, weshalb er neue technische Ideen umsetzen konnte. So erhielt der Typ 32 einen luftgekühlten Vierzylinder-Boxermotor im Heck und anstatt der damals üblichen

Porsche Typ 32, 1933/34 vor der Kronenstraße 24 in Stuttgart, dem damaligen Sitz der Porsche GmbH
Porsche Type 32, a prototype built for NSU, seen in 1933/34 in front of Kronenstrasse 24 in Stuttgart, the Dr. Porsche GmbH company's offices at that time

his technical ideas. The Type 32 was accordingly powered by a rear-mounted flat-four air-cooled engine, and the customary leaf springs were replaced by the "Porsche System" torsion bars that had been patented in 1931. The platform chassis with its central tube and the independent suspension with trailing links at the front and swing axles at the rear were also important

Blattfedern die 1931 patentierte Drehstabfederung »System Porsche«. Auch der Zentralrohr-Plattformrahmen und die Einzelradaufhängung mittels Kurbellenkerachse vorne und Pendelachse hinten nahmen wichtige technische Merkmale des Volkswagens vorweg.

Nach der Genehmigung der Konstruktion durch den NSU-Vorstand Fritz von Falkenhayn am 10. November 1933 entstanden beim Heilbronner Karosseriebauer Drauz zwei Prototypen in Gemischtbauweise mit Holzkarosserie. Für einen dritten Prototypen fertigten die in der Stuttgarter Augustenstraße ansässigen Karosseriewerke Reutter eine Ganzstahlkarosserie an. Insgesamt wirkte der NSU bereits ausgereift und wies deutliche Stilelemente des späteren VW-Käfers auf. Bis zur Serienfertigung schaffte es jedoch auch diese Porsche-Entwicklung nicht. Als die Rezession in Deutschland überwunden geglaubt war und die Motorrad-Konjunktur 1934 wieder anlief, entschieden sich auch die NSU-Werke gegen den Aufbau einer geschätzten zehn Millionen Reichsmark teuren Automobilproduktion.

Die zweite Variante des Typ 32 (2.v.r.) vor der Porsche-Villa 1933/34
The second version of the Type 32 (second from right) in 1933/34, in front of the Porsche villa

technical features that were to reappear on the Volkswagen. After the design had been approved by NSU Director Fritz von Falkenhayn on November 10, 1933, two prototypes of mixed wood and metal construction were built by the Drauz company in Heilbronn.

For a third prototype, the Reutter company, which had its factory on Stuttgart's Augustenstrasse, built an all-steel body. The NSU made a mature impression and exhibited various styling features that would also be seen on the later Volkswagen, but this Porsche design also failed to reach series production status. Although the recession in Germany was considered to be over, and sales of motorcycles picked up again in 1934, NSU decided against investing the estimated ten million Reichsmarks that would have been needed to start automobile production.

DIE PROTOTYPEN DES VOLKSWAGENS
VOLKSWAGEN PROTOTYPES

DIE PROTOTYPEN DES VOLKSWAGENS
VOLKSWAGEN PROTOTYPES

Als der Vertrag zwischen Porsche und dem RDA am 22. Juni 1934 unterzeichnet wurde, bedeutete dies keineswegs den Beginn der Arbeiten am Volkswagen. In Erwartung des Auftrags hatten die Porsche-Konstrukteure schon parallel zu den Typen 12 und 32 im April 1934 mit der Konstruktion des Typ 60 begonnen. Die Realisierung der Konstruktionszeichnungen mit dem Bau der ersten beiden VW-Versuchswagen begann im Frühjahr 1935. Da die Porsche GmbH bislang nur als Konstruktionsbüro tätig gewesen war und alle bisherigen Prototypen unter der Regie der jeweiligen Auftraggeber oder bei Zulieferern gefertigt worden waren, standen keine eigenen Werkstätten zur Verfügung. Die Einrichtung oder Anmietung eines angemessenen Betriebs ließ das knappe Projektbudget nicht zu, weshalb man auf die Doppelgarage

Testfahrt mit den beiden Prototypen V1 (Limousine) und V2 (Cabrio) im Schwarzwald, Ostern 1936
A test run in the Black Forest with the two prototypes V1 (sedan) and V2 (convertible), Easter 1936

Work on the Volkswagen began well before Porsche and the RDA signed their contract on June 22, 1934. Anticipating that the contract would soon be awarded to them, Porsche's designers had begun work on the Type 60 in April 1934, in parallel to the Types 12 and 32. Construction of the first two experimental VW cars on the basis of the design drawings started in the spring of 1935. Until then, Porsche GmbH had only acted as a design office, and previous prototypes had been built either by the client or an outside supplier. The company had no workshops of its own, and the available budget was too limited to set them up or rent them. The only solution was to use the two-car garage at Porsche's private house on Feuerbacher Weg, in the north of Stuttgart. Four years after the opening of the design office, Porsche's team had

des privaten Porsche-Wohnhauses am Feuerbacher Weg in Stuttgart-Nord angewiesen war. Vier Jahre nach Gründung des Konstruktionsbüros war die Porsche-Belegschaft auf 35 Mitarbeiter angestiegen, von denen zwölf Mechaniker für den Bau der Versuchswagen abgestellt wurden. Unter improvisierten und beengten Bedingungen entstanden in der Porsche-Garage bis zur Jahresmitte zwei Prototypen. Die erste fahrfertige Limousine, V1 genannt (V = Versuchswagen), wurde am 3. Juli 1935 einer Kommission des RDA vorgestellt. Der zweite Versuchswagen, ein als V2 bezeichnetes Cabriolet, trat am 22. Dezember seine Jungfernfahrt an. Beide Wagen hatten ein Zentralrohrfahrgestell und Drehstabfederung, die Böden und auch die Karosserien bestanden jedoch noch zum Teil aus einer Holzkonstruktion.

Bei den Motoren herrschte zu diesem Zeitpunkt noch Unklarheit, welches Konzept das am besten geeignete sei. Die Porsche-Techniker entwickelten verschiedene Versionen eines luftgekühlten Antriebs: einen Zweizylinder-Zweitakt-Doppelkolbenmotor, einen Zweizylinder-Zweitakt-Boxermotor, einen Zweizylinder-Boxermotor in Viertaktausführung und schließlich als

Der V1 im Winter 1935/36
The V1 in the winter of 1935/36

Der V2 mit Ferry Porsche am Steuer 1936 in Tübingen
The V2 with Ferry Porsche at the wheel photographed in 1936 in Tübingen

grown to 35 people, twelve of them mechanics with the task of building the experimental cars. Although space was limited and improvisation was the order of the day, two prototypes had been built in Porsche's garage by halfway through the year. V1, the first roadworthy sedan ("V" was the German abbreviation for "Experimental"), was shown to a commission appointed by the RDA on July 3, 1935. The second experimental car (V2), a convertible, was completed and driven by December 22. Both cars had a central-tube chassis and torsion bar suspension. The floor and parts of the body were still made partly of wood at that time.

There was some doubt even then about the most suitable engine concept. Porsche's technical team developed various air-cooled engines: a twin-cylinder, twin-piston two-stroke, a horizontally opposed two-stroke twin, a similar four-stroke unit and, the final version as it turned out, a flat four four-stroke engine. None of the earlier designs performed satisfactorily: power output from the four-stroke flat twin was too low, and the twin-piston two-stroke lacked long-term reliability. Although Ferdinand Porsche at first favored

IIIA-0427

Verschiedene Motorkonzepte wurden ausprobiert. Hier der »C-Motor«, ein 2-Zylinder-Zweitakt-Schiebermotor

Various engine concepts were tried out. This is the "C" engine, a twin-cylinder two-stroke unit with sleeve valves

Der »D-Motor« ist ein
2-Zylinder-Viertakt Motor
The "D" engine was a
four-stroke twin

Linke Seite: Der »D-Motor« war bereits ein Viertakt-Boxer-Motor, allerdings noch mit zwei Zylindern
Left page: Components of the "D" engine, a horizontally opposed four-stroke, though still with only two cylinders

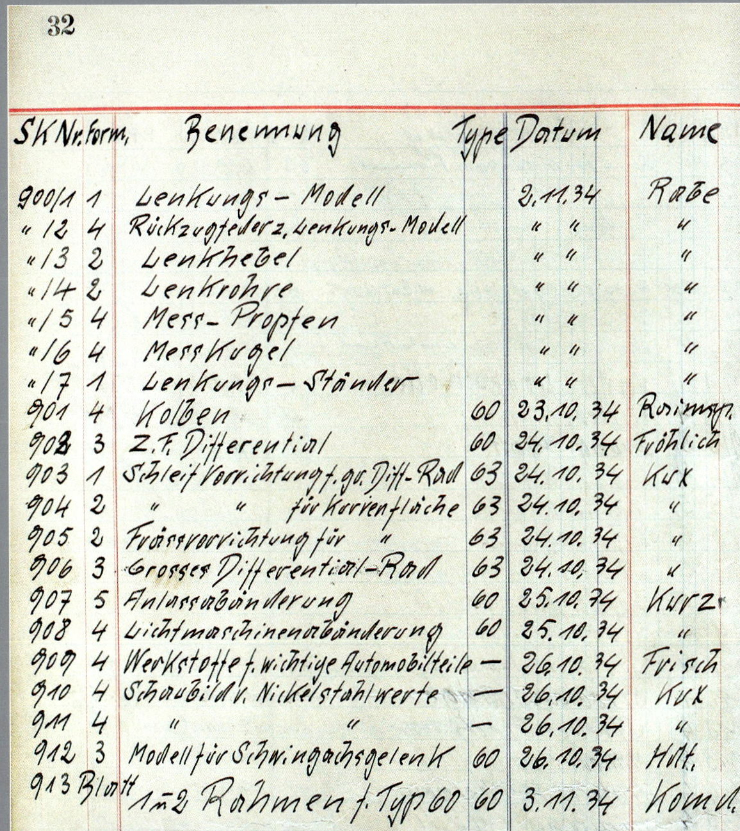

32

SK Nr.	form.	Benennung	Type	Datum	Name
900/11	1	Lenkungs-Modell		3.11.34	Rabe
" 12	4	Rückzugfeder z. Lenkungs-Modell		" "	"
" /13	2	Lenkhebel		" "	"
" /14	2	Lenkrohre		" "	"
" /15	4	Mess-Propfen		" "	"
" /16	4	Messkugel		" "	"
" /17	1	Lenkungs-Ständer		" "	"
901	4	Kolben	60	23.10.34	Reimsp.
902	3	Z.F. Differential	60	24.10.34	Fröhlich
903	1	Schleifvorrichtung f. gr. Diff-Rad	63	24.10.34	Kurz
904	2	" " für Kurvenfläche	63	24.10.34	"
905	2	Fräsvorrichtung für "	63	24.10.34	"
906	3	Grosses Differential-Rad	63	24.10.34	"
907	5	Anlassabänderung	60	25.10.34	Kurz
908	4	Lichtmaschinenabänderung	60	25.10.34	"
909	4	Werkstoffe f. wichtige Automobilteile	—	26.10.34	Frisch
910	4	Schaubild v. Nickelstahlwerte	—	26.10.34	Kurz
911	4	" "	—	26.10.34	"
912	3	Modell für Schwingachsgelenk	60	26.10.34	Holt.
913	Blatt 1 u. 2	Rahmen f. Typ 60	60	3.11.34	Komd.

Dieser Ausschnitt aus der Zeichnungsliste von 1934 zeigt, wie detailliert von vielen Teilen Zeichnungen angefertigt wurden
This extract from the 1934 list of drawings indicates how much detail went into the drawings of many parts

letzte Variante einen Vierzylinder-Viertakt-Boxermotor. Bis zur Fertigstellung des Vierzylinders hatte kein Motor die an ihn gestellten Erwartungen erfüllen können. Beim Viertakt-Zweizylinder mangelte es an Leistung, der Doppelkolben-Motor erbrachte nicht die erforderliche Standfestigkeit. Obwohl Ferdinand Porsche aus Kostengründen lange Zeit einen Zweitaktmotor favorisierte, wurde während der Motorenversuche immer deutlicher, dass allein der vom Porsche-Ingenieur Franz Xaver Reimspiess entworfene Vierzylinder-Boxermotor in der Lage war, allen Anforderungen gerecht zu werden. Obwohl dieser Motor aufgrund seiner vergleichsweise hohen Materialkosten scherzhaft als »Rolls-Royce-Motor« bezeichnet wurde, erwies sich die Reimspiess-Konstruktion letztendlich sogar als rationeller in der Fertigung als die übrigen Entwürfe. Verglichen mit einem Zweizylinder war der Materialaufwand für den Vierzylinder-Boxer zwar höher, aber nicht in dem Maße, wie es seine deutlich höhere Leistungsfähigkeit hätte erwarten lassen.

Nachdem der erste Versuchswagen V1 in der zweiten Jahreshälfte 1935 ausführlich getestet worden war, wurde der Prototyp zusammen mit dem zweiten Versuchswagen V2 am 29. Dezember 1935 auf eigener Achse nach München überführt. An Bord der beiden Fahrzeuge befanden sich Ferdinand Porsche, Sohn Ferry, Schwiegersohn Anton Piëch, Chauffeur Josef Goldinger und zwei Mechaniker. In der bayrischen Landeshauptstadt wurden die Fahrzeuge in einer einstündigen Präsentation Adolf Hitler vorgeführt, worauf die Vertreter des RDA erbost reagierten, da sie vorher nicht informiert worden waren.

a two-stroke engine for cost reasons, engine tests made it increasingly clear that the only engine capable of satisfying every requirement was the flat four four-stroke unit developed by Porsche engineer Franz Xaver Reimspiess. It incurred relatively high material costs, and for this reason was sometimes mocked as the "Rolls-Royce engine", but in the final analysis, the Reimspiess design proved to be easier to build cost-effectively than the others. A four-cylinder horizontally opposed four-stroke naturally contained more components than a two-stroke engine, but this drawback was more than made good by its higher performance.

After thorough testing of V1, the first experimental car, in the second half of 1935, this prototype and V2, the second car, were driven under their own power to Munich on December 29, 1935. Ferdinand Porsche, his son Ferry, his son-in-law Anton Piëch, company driver Josef Goldinger and two mechanics traveled in the cars. On arrival in Bavaria's capital city, the cars were demonstrated to Adolf Hitler in a presentation lasting an hour; this angered the RDA's representatives, who had not been informed in advance.

Linke Seite: Der V1 (hinten) während einer Testfahrt, 1935/36
Left page: The V1 (at the back) during a test run in 1935/36

Ferdinand Porsche 1935
Ferdinand Porsche 1935

Das Fahrwerk des ersten Prototyps. Der Boden bestand noch aus Holz, von einem Metall-Zentralrohrrahmen getragen
The chassis of the first prototype. It had a wooden floor supported on a metal central-tube frame

Die beiden Prototypen V1 und V2 wurden in der Garage der Porsche-Villa gebaut
The two prototypes, V1 and V2, were built in the garage of the Porsche villa

DAUERTEST: DIE VERSUCHSWAGENREIHE V3

Während im Februar 1936 mit dem Bau drei weiterer Volkswagen-Prototypen mit der Bezeichnung V3 begonnen wurde, kam seitens des »Reichsverband der Automobilindustrie« vermehrt Widerstand auf. Terminverzögerungen und Kostenüberschreitungen sowie technische Probleme der Versuchsfahrzeuge führten zu heftiger Kritik des RDA-Präsidenten Robert Allmers, als die Testwagen V1 und V2 den Vertretern der Automobilindustrie am 24. Februar 1936 in Berlin präsentiert wurden. Unbeeinträchtigt davon führte Ferdinand Porsche die Arbeiten an einer dritten Versuchswagenserie weiter, die er bereits einige Monate zuvor vom RDA hatte genehmigen lassen. Wie schon die V1- und V2-Wagen entstanden auch die drei V3-Exemplare in der Garage der Porsche-Villa, aus der das erste Fahrzeug am 6. Juni 1936 seine Jungfernfahrt antrat. Mit Ausnahme der Karosserieform wich die mit Vierzylinder-Boxermotor und Ganzstahlchassis ausgestattete V3-Serie des Volkswagens nur noch in Details von der späteren Serienversion ab. Nachdem zwei der V3-Wagen am 11. Juli 1936 erneut mit positivem Ergebnis vorgeführt wurden, begann

Die nächste Prototypenreihe hieß »V3«. Drei leicht verschiedene Prototypen wurden gebaut
The next series of prototypes was known as "V3". Three cars with slight differences were built

LONG-TERM TESTING: THE V3 SERIES OF EXPERIMENTAL CARS

In February 1936, work began on building three more Volkswagen prototypes; these were designated "V3". At the same time, resistance from the RDA, the German Automobile Industry Association, became more severe. Deadlines were not always met, cost targets were exceeded and the experimental cars suffered certain technical problems. When prototypes V1 and V2 were shown to automobile industry representatives in Berlin on February 24, 1936, there was violent criticism from the RDA's President, Robert Allmers. Ferdinand Porsche ignored this and continued work on the third series of experimental cars, for which he had obtained the RDA's approval some months earlier. Like V1 and V2, the three V3 cars were built in the garage of Porsche's house: the first of them was ready to drive on June 6, 1936. Apart from the shape of the body, the V3 series with its flat-four engine and all-steel chassis differed only in detail from the later production version. On July 11, 1936 two of the V3 cars were demonstrated again successfully, and systematic road testing began. This took the form

IIIA-0427

die systematische Fahrerprobung im Rahmen einer vom RDA kontrollierten Dauerversuchsfahrt. Dabei forderte der Reichsverband, dass jeder der drei Wagen 30.000 Kilometer zurücklegen musste, was bald darauf auf eine Distanz von 50.000 Kilometer erhöht wurde. Während eines bis zu 15 Stunden langen Testtages absolvierten Fahrzeug und Fahrer Strecken von fast 800 Kilometern. Das am 12. Oktober 1936 begonnene und von Ferry Porsche geleitete Fahrprogramm sah täglich zwei verschiedene Routen vor: Vormittags eine 360 Kilometer lange Runde über den Schwarzwald, nachmittags 440 Kilometer Autobahn bis Bad Nauheim und zurück.

Eine Fahrerprobung dieses Umfanges war ein Novum innerhalb der Automobilindustrie, denn kein anderer Fahrzeugtyp war je zuvor einem solch harten Test unterzogen worden. Neben den Testfahrern befanden sich in den Wagen junge Ingenieure der Technischen Hochschulen Stuttgart und Berlin, die im Auftrag des RDA alle Vorkommnisse genau protokollierten. Zusätzlich saß noch ein Vertreter der Firma Porsche dabei, weshalb die Wagen oft mit drei, manchmal mit vier Personen besetzt waren. Dass die Erprobung

Ein 1:5 Modell des V3 im Garten der Porsche-Villa
A one-fifth scale model of the V3 in the garden of the Porsche villa

Die ersten beiden V3 hatten noch keine in die Kotflügel integrierten Scheinwerfer
The first two V3 cars did not yet have the headlamps on the front fenders

of a long-term program supervised by the RDA. The manufacturers' association insisted at first that each of the three cars should complete a distance of 30,000 kilometers, but later increased this figure to 50,000 km. Testing continued for up to 15 hours a day, with the cars and their drivers covering almost 800 kilometers. Begun on October 12, 1936 and organized by Ferry Porsche, the road test program was carried out on two different routes every day. In the morning, the cars were driven through the Black Forest for 360 kilometers; in the afternoon they took to Germany's 'autobahn' highways and were driven as far as Bad Nauheim and back.

Such an extensive road test schedule was new to the automobile industry; until then, no other new model had been subjected to such tough treatment. Young engineering students from the Colleges of Advanced Technology in Stuttgart and Berlin rode in the cars on behalf of the RDA, and kept a careful record of their performance and all unusual occurrences. A Porsche company representative was also on board, which meant that the cars normally had three and sometimes even four occupants. To make the task even tougher,

IIIA - 0428

Die V3 hatten als erste Prototypen Vierzylinder-Viertakt-Boxermotoren, hier in der Porsche-Garage während der Montage
The V3s were the first prototypes to have flat-four four-stroke engines; they were assembled here in Porsche's garage

Rechte Seite: Der V3/3, der dritte Wagen aus der V3-Serie, erhielt erstmals eine Stahlkarosserie
Right page: The V3/3, the third car in the V3 series, was the first with an all-steel body.

IIA-34993

Von Oktober bis Dezember 1936 fand eine Fahrerprobung statt, während der jeder der drei V3 mehr als 50.000 km zurücklegen sollte. Hier die drei Wagen auf dem Hof der Porsche-Villa

From October to December 1936 the V3 cars underwent a series of test runs in which each of them was expected to cover 50,000 kilometers. The three cars are seen here at the Porsche villa

Der V3/1 1936 in der
Einfahrt zur Porsche-Villa
*The V3/1 in the entrance
to the Porsche villa, 1936*

Rechte Seite: Eine Testfahrt
führte die Prototypen V3/1
(hinten) und V3/3 im September 1936 über die Alpen
*Right page: Driving the V3/1
(at the rear) and V3/3 in the
Alps, September 1936*

IIIA-34993

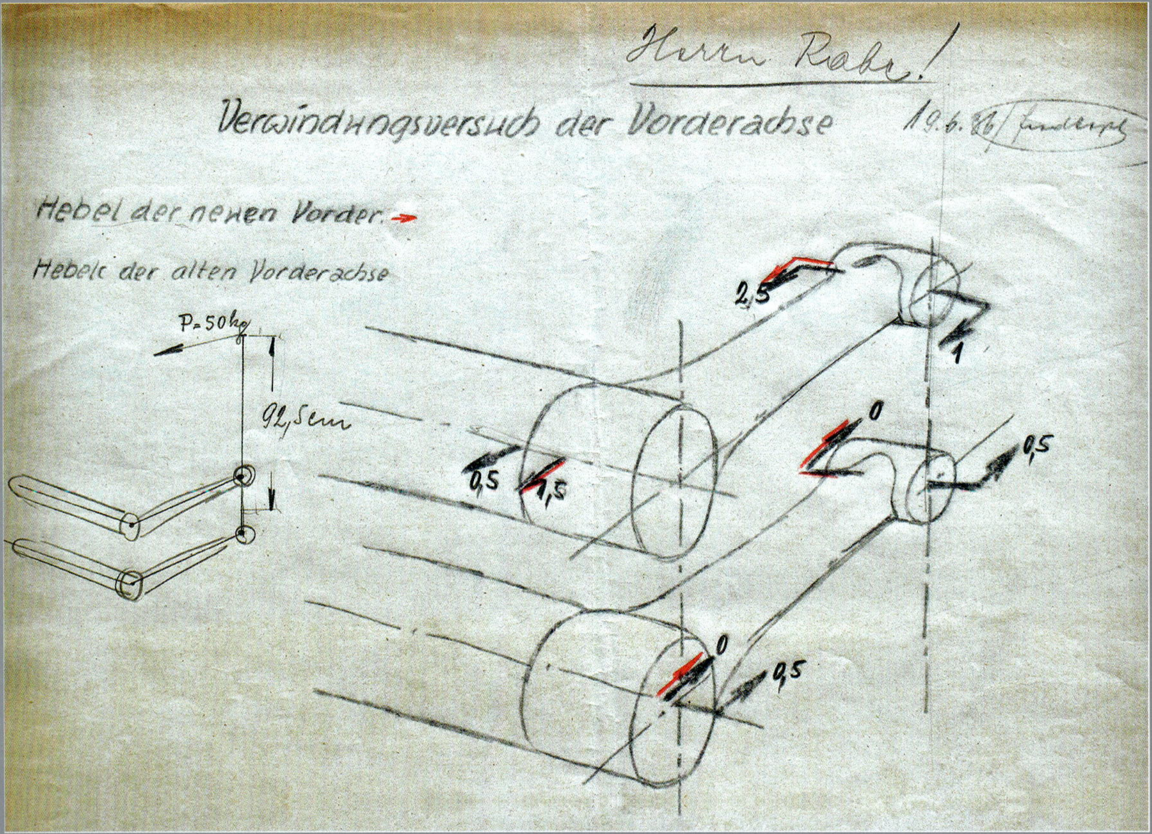

Die Porsche-Drehstabfederung wurde auch für den Typ 60 eingesetzt
Torsion-bar suspension, a Porsche invention, was of course used on the Type 60 as well

Rechte Seite: Die V3-Testfahrt begann vormittags mit dieser Rundfahrt durch den Schwarzwald, nachmittags ging es über die Autobahn nach Bad Nauheim und zurück
Right page: The V3 test run began in the morning on this route through the Black Forest, then to Bad Nauheim and back in the afternoon, mostly on the 'autobahn'

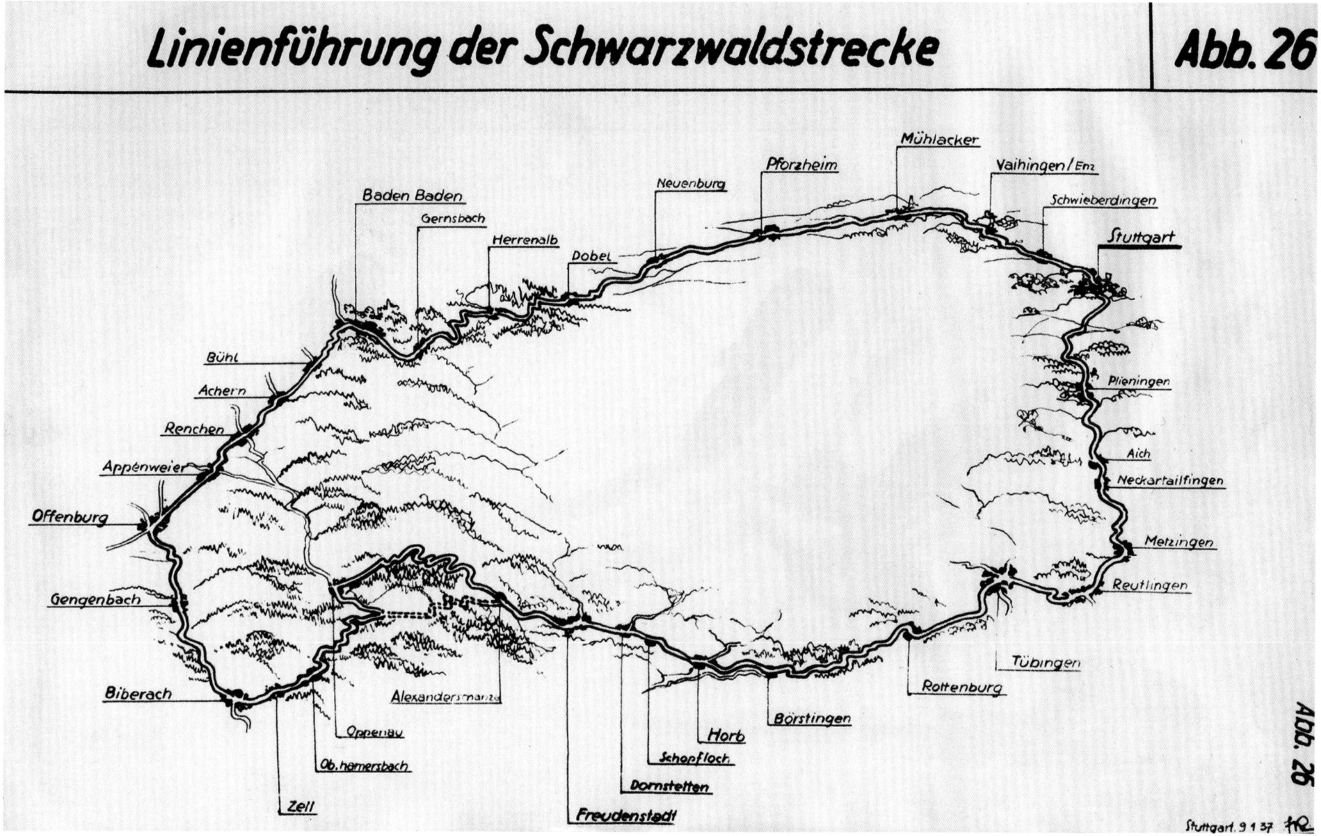

Linienführung der Schwarzwaldstrecke
Abb. 26
Mühlacker
Pforzheim
Vaihingen/Enz
Neuenburg
Schwieberdingen
Baden Baden
Gernsbach
Herrenalb
Dobel
Stuttgart
Bühl
Achern
Renchen
Appenweier
Offenburg
Plieningen
Aich
Neckartailfingen
Metzingen
Reutlingen
Gengenbach
Biberach
Alexandersmanze
Tübingen
Rottenburg
Börstingen
Horb
Schopfloch
Oppenau
Ob. hamersbach
Dornstetten
Zell
Freudenstadt
Abb. 26
Stuttgart. 9 1 37

aus Zeitmangel im Herbst und Winter durchgeführt wurde, kam erschwerend hinzu. Am 22. Dezember 1936 endete die Fahrerprobung. Hinter den Testwagen lagen je 50.000 Kilometer, ein paar Unfälle und verschiedene Materialschäden. Diese waren jedoch nicht auf grundsätzliche Fehler des Fahrzeugs, sondern auf die behelfsmäßige Herstellung der Fahrzeuge zurückzuführen. In seinem Abschlussbericht vom 26. Januar 1937 musste der RDA feststellen, dass »die Bauart sich als zweckmäßig erwies« und die Fahrzeugeigenschaften »eine Weiterentwicklung empfehlenswert erscheinen lassen.«

Links: Die spartanische Innenaustattung des V3 zeigt, dass es sich noch um reine Prototypen handelte
Left: The V3's very sparse interior equipment and trim tells us that this is still a prototype

Rechts: Die V3 hatten noch kein Heckfenster, da die Luftschlitze zur Motorkühlung nötig waren und kein Platz für ein Heckfenster ließen
Right: The V3 was built without a rear window: the space was needed for the engine's air cooling vents

shortage of time led to the tests being carried out in the fall and winter. The test program was completed on December 22, 1936. Each of the cars had covered the required 50,000 kilometers; there had been a few accidents and various material failures, but these were not due to fundamental flaws in the cars' design but to the provisional methods used to build the prototypes. In its concluding report, issued on January 26, 1937, the RDA was obliged to concede that "the design has proved practicable" and their performance "recommends them for further development".

Das Fahrgestell des V3 wurde schon komplett aus Stahl gefertigt, allerdings ist hier zwischendurch noch der 2-Zylinder-Viertakt Motor eingebaut

The V3's chassis was entirely made of steel; this car is still powered by the four-stroke flat twin engine

Der V3/1 hatte eine Karosserie aus Metall-Holz-Mischbauweise. Die Windschutzscheibe bekam nachträglich Heizdrähte, um die Sicht im Herbst 1936 zu verbessern
The V3/1 still had a combined wood and metal body. In the fall of 1936 a windshield with wire heating elements was fitted to give the driver a better view in poor weather

Rechte Seite: Während der 50.000-Kilometer-Testfahrt wurden möglichst viele Kilometer auf der Autobahn zurückgelegt
Right page: As much as possible of the 50,000-kilometer test run was performed on the new German 'autobahn' highway

IIIA-34992
D

DIE VOLKSWAGEN W30-SERIE

Um weitere Erkenntnisse für eine Serienproduktion gewinnen zu können, wurde eine neue Serie von VW-Prototypen geplant, die in einer Automobilfabrik anstelle der improvisierten Porsche-Garage montiert werden sollten. Noch während der Vorbereitung der V3-Dauererprobung hatte der RDA bei der Daimler-Benz AG den Bau einer Kleinserie von 30 VW-Prototypen angefragt, was Daimler-Vorstand Wilhelm Kissel am 23. Juni 1936 bestätigte.

Unter der Aufsicht von Porsche-Ingenieuren wurden am 9. September 1936 im Daimler-Werk die Arbeiten an der Volkswagen-Serie W30 aufgenommen. Bis zum 15. Mai 1937 waren insgesamt 30 Wagen entstanden, darunter ein Cabriolet. Eine Ausnahme war der erste produzierte Wagen mit dem Kennzeichen IIIA-37000 (IIIA stand bis 1945 für Stuttgart). Dieser entstand noch in der Porsche-Villa und war bereits parallel zum Abschlussbericht der V3-Testfahrt im Januar 1937 fertig gestellt worden. Die übrigen 29 Versuchswagen wurden wie geplant bei Daimler-Benz gebaut, wobei die Motoren und Fahrgestelle in Untertürkheim

Die 30 Wagen der Serie W30 entstanden bei Daimler-Benz, nur der erste W30 wurde in der Garage der Porsche-Villa gebaut
The thirty cars in the W30 series were built at the Daimler-Benz plant, except for the first car, which was assembled in the garage at the Porsche villa

THE VOLKSWAGEN W30 SERIES

The next step was to gather information relevant to volume production, and so a new series of VW prototypes was planned, not to be built in Porsche's improvised garage workshop this time but on a car factory's assembly line. Before the long-term V3 tests had been carried out, the RDA had asked Daimler-Benz AG whether it could build a small batch of 30 Volkswagen prototypes, and on June 23, 1936 Daimler's Board Chairman Wilhelm Kissel agreed to this.

Supervised by Porsche engineers, work on the W30 series of Volkswagens began at Daimler's factory on September 9, 1936. The thirty cars, including a convertible, were completed by May 15, 1937. The first car in this series, however, had been built at the Porsche villa and was already complete by the time the concluding report on the road test program was issued in January 1937. It bore the license plate IIIA-37000 (IIIA was allocated to Stuttgart in the system used until 1945). The remaining 29 prototypes were built by Daimler-Benz: the engines and chassis in Untertürkheim, the bodies at Daimler's car body

IIIA
37000

gefertigt wurden, die Aufbauten jedoch im Daimler-Karosseriewerk in Sindelfingen entstanden waren. Technisch basierten die W30 auf dem Ganzstahlwagen der V3-Serie, unterschieden von diesem aber in zahlreichen Details. Die Karosserie war abermals verändert worden, das Fahrgestell erhielt zur Erhöhung der Steifigkeit an der Unterseite des Zentralrohrs eingeprägte Sicken. Im Gegensatz zum ringförmigen Ölkühler des V3-Motors verfügten die W30-Aggregate über stehende Ölkühler. Daneben wurde die elektrische Benzinpumpe, die beim V3 für Schwierigkeiten gesorgt hatte, durch eine mechanische Kraftstoffförderung ersetzt.

Auch die Wagen der W30-Serie sollten einer ausführlichen Fahrerprobung unterzogen werden, doch zuvor kam es im Frühjahr 1937 zu einer wesentlichen organisatorischen Änderung. Aufgrund des stetigen Widerstands des RDA gegen den Volkswagen wurde dem Reichsverband im März 1937 die Projektleitung entzogen. In den Vertrag des RDA trat nun der Einheitsverband der Arbeitnehmer und Arbeitgeber ein, die »Deutsche Arbeitsfront« (DAF). Die 1933 durch die Übernahme der freien Gewerkschaften gegründete

Der erste W30 (links) und der V3/3 im Winter 1936/37
The first W30 (left) and the V3/3, in the winter of 1936/37

Rechte Seite: Das erste Fahrzeug der W30-Serie
Right page: The first car of the W30 series

construction plant in Sindelfingen. The W30 was based technically on the all-steel V3 car, but with many detail modifications. The body underwent further changes and chassis rigidity was improved by forming swage-lines on the underside of the central tube. The V3's ring-shaped oil cooler was dropped in favor of a vertical unit on the W30 engine, and the electric fuel pump, which had caused problems on the V3, was replaced by a mechanical unit.

Extensive road testing was also planned for the W30 cars, but in the spring of 1937, before this program could be started, a significant organizational change took place. The RDA's unceasing hostility to the project led to it being taken away from them in March 1937. The Association's contract was transferred to the "Deutsche Arbeitsfront" (DAF), Germany's united employers' and workers' organization, which had been formed in 1933 to absorb the independent trade unions. It possessed sufficient financial resources to tackle even a large-scale project such as the "German People's Car". On May 28, 1937 a "People's Car Preparation Company" was formed and the project's funds guaranteed with the DAF's capital. This

Linke Seite: Ferdinand Porsche (links) verfolgt die Montage des W30 bei Daimler-Benz in Untertürkheim
Left page: Ferdinand Porsche (left) supervising assembly of the W30 at Daimler-Benz in Untertürkheim

Der Motor des W30 war der bereits bei den V3-Prototypen getestete luftgekühlte Vierzylinder-Viertakt-Boxermotor
The W30's engine was the air-cooled flat four unit already tested in the V3 prototypes

Organisation verfügte über ausreichende finanzielle Mittel um ein Großprojekt wie den deutschen Volkswagen zu realisieren. Am 28. Mai 1937 wurde die »Gesellschaft zur Vorbereitung des Volkswagens« (GezuVor) gegründet und mit Kapital der DAF ausgestattet, so dass auch die Finanzierung des Projekts endgültig gesichert war. Als Geschäftsführer der Gesellschaft wurden Ferdinand Porsche, Jacob Werlin sowie Bodo Lafferentz ernannt, der innerhalb der DAF ab 1938 auch Leiter der Organisation »Kraft durch Freude« (KdF) wurde. Ferdinand Porsche war neben der Fahrzeugentwicklung nun auch offiziell für die Technik und Planung eines zukünftigen Volkswagenwerks zuständig.

Die Entscheidung zum Bau einer eigenen Volkswagen-Fabrik war nicht unumstritten; noch 1936 wurde diskutiert, den VW bei den verschiedenen Mitgliedsfirmen des RDA produzieren zu lassen. Den Ausschlag für ein eigenes VW-Werk gaben, neben den Schwierigkeiten mit dem RDA, auch die Finanzierungsprobleme des Volkswagen-Projektes. Um den Wagen zum politisch geforderten Verkaufspreis von 990 Reichsmark produzieren zu können, waren riesige Stückzahlen und

Das Fahrgestell des ersten W30, fotografiert während des 50.000 km Dauerversuchs der V3-Prototypen
The chassis of the first W30, photographed while the 50,000 km long-term test with the V3 prototypes was still in progress

preparation company was known by its German acronym “GezuVor”, a play on words with the convenient meaning of “go ahead”. Ferdinand Porsche, Jacob Werlin and Bodo Lafferentz were appointed its general managers, the last-named taking charge within the DAF from 1938 on of the organization known as “Kraft durch Freude” (KdF), which translates as “Strength through Joy”. In addition to his vehicle development work, Ferdinand Porsche was now officially responsible for the technical features and planning of the plant that was to be built to produce the Volkswagen.

The decision to build a new Volkswagen manufacturing plant did not go undisputed. Even in 1936, there was an alternative plan to have the car built at various RDA member-companies' plants. That an independent VW production facility should be built was not solely due to the RDA's obstructive attitude but also to problems associated with financing the project. To sell the car at the target price of 990 Reichsmarks laid down by the politicians, it would have to be built in vast numbers and the latest production methods adopted, despite the fact that these came from America.

IIIA
34882

Linke Seite: Heckansicht des W30-Fahrgestells
Left page: The W30 chassis with engine

Das Interieur des W30 war zwar spartanisch, jedoch im Vergleich zu den V3-Prototypen deutlich verbessert
Although the W30 still had a very spartan interior, it was already a great improvement on the V3 prototypes

die neuesten, wohlgemerkt amerikanischen, Produktionsmethoden (wie das in Deutschland damals noch kaum verwendete Punktschweißverfahren) notwendig. Beides ließ sich nicht in den bestehenden Automobilfabriken verwirklichen. Um die fortschrittlichen Methoden des amerikanischen Automobilbaus zu studieren, hatte Ferdinand Porsche bereits im Herbst 1936 eine sechswöchige Studienreise in die Vereinigten Staaten von Amerika unternommen. In Detroit besichtigte er die Produktionsanlagen der Ford Motor Company und der General Motors Corporation. Im Juni 1937 folgte eine zweite Studienreise in die USA. In Begleitung seines Sohnes Ferry und den zwei anderen GezuVor-Geschäftsführern Bodo Lafferentz und Jakob Werlin war der Zweck der Reise neben dem Ankauf von Spezialmaschinen auch die Anwerbung von nach Amerika ausgewanderten Produktionsexperten, die für das künftige Volkswagenwerk gewonnen werden sollten.

Die Fahrerprobung der Volkswagen W30-Serie wurde 1937 noch in weit aufwändigerem Rahmen durchgeführt, als es im Vorjahr mit den V3-Wagen der Fall war. Als Ausgangsort der Testfahrten diente nicht mehr die Porsche-Villa, sondern eine Panzerkaserne im nahe bei

Die Türen des W30 waren noch an der B-Säule montiert, was später aus Sicherheitsgründen verworfen wurde

The W30's doors were at first hinged to the B posts of the body, but this was later modified for safety reasons

A typical example was spot welding of the car bodies, a method scarcely known in Germany at that time. RDA members' existing factories were unable to handle this major increase in capacity, or to adopt the new production methods without serious upheavals. In the fall of 1936 Ferdinand Porsche had undertaken a six-week study trip to the United States of America. In Detroit he toured the Ford Motor Company's and General Motors Corporation's production facilities. On a second such visit in June 1937, he was accompanied by his son Ferry and the two other 'GezuVor' general managers, Bodo Lafferentz and Jakob Werlin. This trip was devoted to the purchase of special-purpose machinery, but also to headhunting production experts who had emigrated in earlier years, and who were now needed for the projected Volkswagen plant.

Road tests of the Volkswagen W30 series in 1937 were conducted on a much larger scale than with the V3 cars a year before. Instead of using the Porsche villa as a starting point, the test team was moved to an army tank barracks in Kornwestheim, a town not far from Zuffenhausen, where an SS vehicle squad was based. Ferdinand Porsche's son Ferry supervised

Zuffenhausen gelegenen Kornwestheim, in der eine SS-Fahrbereitschafts-Staffel untergebracht wurde. Als Leiter der Versuchsfahrten wurde Ferdinand Porsches Sohn Ferry eingesetzt, der von Albert Liese, der bereits an der V3-Testfahrt mitgewirkt hatte, unterstützt wurde. Die Tests begannen am 15. Mai 1937 und führten die Versuchswagen über eine Gesamtdistanz von 2,4 Millionen Versuchskilometern, wobei einzelne Wagen Laufleistungen von mehr als 100.000 Kilometer absolvierten. Kontinuierlich und unter strengster Geheimhaltung arbeiteten die Techniker daran, vor allem die Standfestigkeit der Wagen zu erhöhen. Vom 10. bis 16. September nahmen die W30-Wagen sogar an einer Alpenfahrt teil, was in der Folge immer wieder Bestandteil der Erprobungen sein sollte. Besonders im Jahr 1938 waren die Gebirgsstrecken der Alpen ein häufig angesteuertes Erprobungsfeld der verschiedenen Volkswagen-Prototypen.

Das Rückfenster lag zunächst zwischen dem Innenraum und dem Motorraum. Durch die Kühlluftschlitze in der Motorhaube konnte der Fahrer nach hinten sehen
At first there was no rear window in the engine cover, only between the car's interior and the engine compartment, but a partial view to the rear was possible through the engine's air cooling vents

the test program, aided by Albert Liese, who had previously helped with V3 testing. The new program started on May 15, 1937: altogether, the cars were to cover 2,400,000 kilometers, with individual cars often clocking up more than 100,000 km. The technical staff toiled unceasingly, in conditions of the utmost secrecy, to make the cars even more reliable. From September 10 to 16 the W30 cars were taken to the Alps, an aspect of their testing that was often repeated later. In 1938 in particular, Alpine mountain passes were regularly used for testing the various Volkswagen prototypes.

IIIA-37015
IIIA 37006

Linke Seite: Zwei W30 während einer der zahlreichen Testfahrten 1937
Left page: Two W30s during one of the many test runs carried out in 1937

Ein Fahrgestell (links), das einzige Cabriolet (mitte) und die Limousinen-Variante (rechts) der W30-Serie
From left to right, the chassis, the only W30 pre-production convertible and the sedan

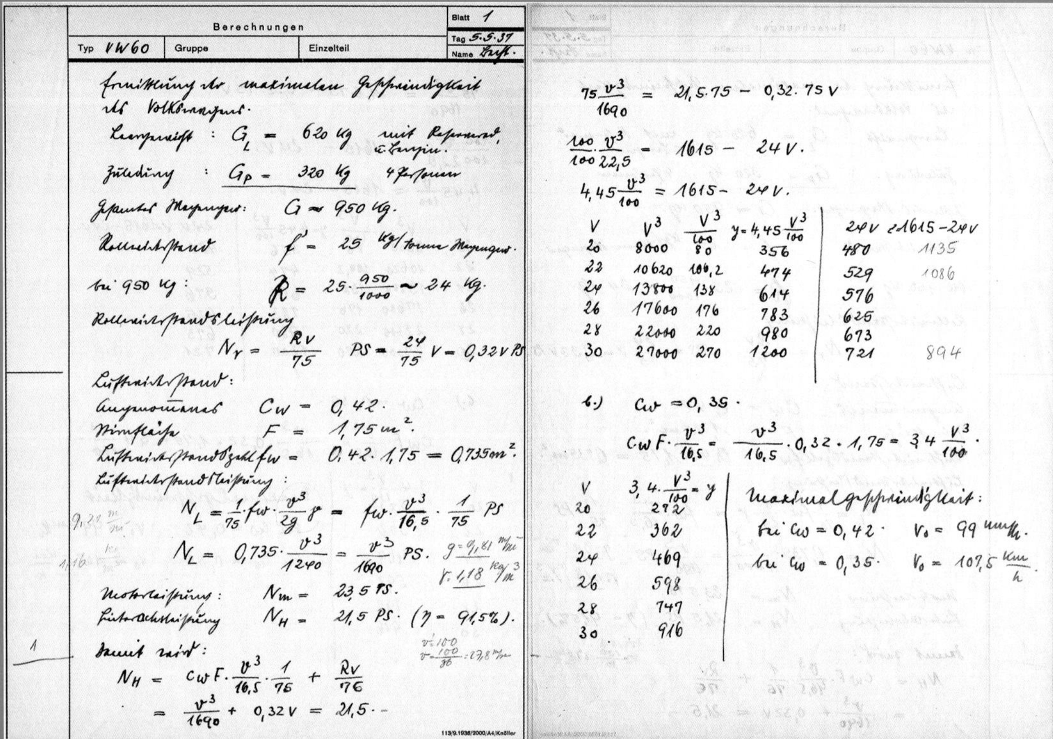
Berechnungen

Blatt 1 | Tag 5.5.37 | Name ...

Typ VW60 | Gruppe | Einzelteil

Ermittlung der maximalen Geschwindigkeit des Volkswagens.

Leergewicht : $G_L = 620$ kg mit Reserverad, u. Benzin.

Zuladung : $G_P = 320$ kg 4 Personen

Gesamtes Wagengew.: $G \approx 950$ kg.

Rollwiderstand $f = 25$ kg/tonne Wagengew.

bei 950 kg : $R = 25 \cdot \frac{950}{1000} \approx 24$ kg.

Rollwiderstandsleistung

$N_r = \frac{Rv}{75}$ PS $= \frac{24}{75} v = 0{,}32 v$ PS

Luftwiderstand:

Angenommenes $C_W = 0{,}42$.

Stirnfläche $F = 1{,}75\ m^2$.

Luftwiderstandszahl $f_W = 0{,}42 \cdot 1{,}75 = 0{,}735\ m^2$.

Luftwiderstandsleistung:

$N_L = \frac{1}{75} f_W \cdot \frac{v^3}{2g} \gamma = f_W \cdot \frac{v^3}{16{,}5} \cdot \frac{1}{75}$ PS

$N_L = 0{,}735 \cdot \frac{v^3}{1240} = \frac{v^3}{1690}$ PS. $g = 9{,}81\ m/s^2$, $\gamma = 1{,}18\ kg/m^3$

Motorleistung: $N_m = 23{,}5$ PS.

Hinterachsleistung $N_H = 21{,}5$ PS. $(\eta = 91{,}5\%)$.

Damit wird:

$N_H = c_W F \cdot \frac{v^3}{16{,}5} \cdot \frac{1}{75} + \frac{Rv}{75}$

$= \frac{v^3}{1690} + 0{,}32 v = 21{,}5$

113/9.1936/2000/A4/Knöller

$75 \frac{v^3}{1690} = 21{,}5 \cdot 75 - 0{,}32 \cdot 75 v$

$\frac{100 \cdot v^3}{100 \cdot 22{,}5} = 1615 - 24 v$.

$4{,}45 \frac{v^3}{100} = 1615 - 24 v$.

V	V^3	$\frac{V^3}{100}$	$y = 4{,}45 \frac{V^3}{100}$	24 V	$z = 1615 - 24V$
20	8000	80	356	480	1135
22	10620	106,2	474	529	1086
24	13800	138	614	576	
26	17600	176	783	625	
28	22000	220	980	673	
30	27000	270	1200	721	894

b.) $c_W = 0{,}35$.

$c_W F \cdot \frac{v^3}{16{,}5} = \frac{v^3}{16{,}5} \cdot 0{,}32 \cdot 1{,}75 = 3{,}4 \frac{v^3}{100}$.

V	$3{,}4 \cdot \frac{V^3}{100} = y$
20	272
22	362
24	469
26	598
28	747
30	916

Maximalgeschwindigkeit:

bei $c_W = 0{,}42$: $V_0 = 99$ km/h.

bei $c_W = 0{,}35$: $V_0 = 107{,}5$ km/h.

Die Berechnung der Höchstgeschwindigkeit des Wagens war abhängig von Gewicht, Rollwiderstand, Motorleistung sowie Stirnfläche und Luftwiderstand und ergab 99 km/h bei C_w 0,42 (107,5 km/h bei C_w 0,35)

Calculation of the car's top speed from its weight, rolling resistance, power output and aerodynamic drag coefficient (c_D) yielded figures of 99 km/h for a c_D of 0.42 and 107.5 km/h for a c_D of 0.35

Rechte Seite: Die W30-Serie aufgereiht in einer Kaserne in Kornwestheim für die Testfahrten 1937

Right page: W30 cars lined up in a military barracks in Kornwestheim, ready for test runs

Senkung
VL
VR
Federung

Linke Seite: Cockpit des W30 während der Testfahrten 1937
Left page: The interior of the W30 as used for test runs in 1937

Reparatur der hinteren Trommelbremse in den Alpen 1937
Repairing the rear drum-brakes in the Alps, 1937

Alpenversuchsfahrt mit der W30-Vorserie 1937, vorne das einzige W30-Cabriolet
Test driving in the Alps with W30 pre-production cars; the only W30 convertible is at the front

Blick aus dem W30 während einer Versuchsfahrt
The W30's interior, photographed during a test run

FIAT
LUBRIFICANTI
P.N.P. • 38 • STELVIO • A.A.S.S
S. CATERINA • 15
STELVIO • 22
R. A. C. I.
IIIA 42803
IIIA-37022

Linke Seite: Alpenversuchsfahrt im Juli 1938 u.a. mit dem W30-Cabriolet (Mitte), links Ferry Porsche
Left page: Test driving in the Alps in July 1938, including the W30 convertible (center); Ferry Porsche is at the left

Ein W30 in der Einfahrt des Porsche Werk 1 in Stuttgart-Zuffenhausen 1938
A W30 enters Porsche's #1 plant in Stuttgart Zuffenhausen, 1938

Linke Seite: Der Innenhof des Porsche-Werk 1 mit W30-Versuchswagen, 1938
Left page: Inner yard of Porsche's #1 plant in 1938, with a W30 car

1942 wurden die Wagen aus der W30-Vorserie verschrottet
Cars from the W30 pre-production series were scrapped in 1942

DIE ERSTEN »BREZELKÄFER«

Obwohl die Ergebnisse des »Massentests« mit den dreißig Versuchswagen positiv verlaufen waren, hatten auch die W30-Prototypen noch keinen Serienstandard erreicht. Ab dem Sommer 1938 folgte eine weitere Version, von der bei der Firma Reutter zunächst ein Holzmodell im Maßstab 1:1 gefertigt wurde. Als fahrtüchtige Prototypen entstanden daraufhin drei Exemplare: Als Variante V303, im Prinzip ein verbesserter W30, entstanden eine Limousine, ein Cabriolet und eine so genannte »Cabriolimousine«; dies war ein geschlossener Wagen mit dem in den 1930er Jahren beliebten Rolldach. Die Verbesserungen betrafen vor allem die Karosserie, die nun ihre beinahe endgültige Form erreichte. Wichtiges Merkmal war das charakteristische zweigeteilte »Brezel-Heckfenster«, das erst durch die Verwendung eines neuen Kühlluftgebläses möglich geworden war. Hierdurch konnte auf einen Teil der Belüftungsschlitze in der Motorhaube verzichtet werden, wodurch Platz für ein Rückfenster geschaffen wurde. Das zweigeteilte Ovalfenster sollte charakteristisch für den Volkswagen werden und ging unter dem Begriff »Brezelfenster« in die

Holzmodell des V303 in Zuffenhausen bei der Karosseriefabrik Reutter (1938)
Wooden model of the V303 at the Zuffenhausen plant of Reutter, the body supplier, in 1938

V303: THE FIRST "PRETZEL BEETLES"

Although results from the large-scale road test program with no fewer than 30 prototype cars were favorable, the W30 had not yet reached volume-production maturity. A further version was produced, starting in the summer of 1938, initially as a full-scale wooden mock-up built by the Reutter company. After this, three roadgoing cars were constructed and known as the V303. They were in effect improved versions of the W30. One was a sedan, the second a convertible and the third a sedan with roll-back roof, a body style that was very popular in the 1930s. Most of the improvements concerned the body, which was now approaching its final form. One of the major changes was the two-part "pretzel" rear window. Space for this had been found by adopting a new engine cooling fan design that dispensed with some of the cooling air vents in the engine hood. This centrally divided oval rear window was to become one of the characteristic features of early Volkswagens. The term "pretzel" came from the window's shape, which resembled the typical loops of this popular bakery product and tied in well with the car's German origin.

IIIA - 39063

Automobilgeschichte ein. Daneben waren bei dieser verbesserten V303-Karosserie die Türen erstmalig an der A-Säule angeschlagen. Dies war vom Chef-Stilisten des Volkswagens, dem Leiter der Porsche-Karosserieentwicklung Erwin Komenda, schon Anfang 1936 vorgeschlagen worden, wurde aber erst umgesetzt, nachdem Ferdinand Porsche auf einer USA-Reise feststellt hatte, dass so gut wie alle neuen Wagen in den USA dieses Merkmal aufwiesen. Als eine weitere Verbesserung waren am V303 Stoßstangen montiert, welche beim W30 nur versuchsmäßig an einzelnen Wagen angebracht wurden. Die Kofferraumhaube der V303 war zwecks einfacherer Herstellung weiter heruntergezogen, was nebenbei dem Erscheinungsbild des Wagens zugute kam. Die Karosserien der Kleinserie wurden nicht wie bisher bei Daimler-Benz gebaut, sondern bei der Stuttgarter Karosseriefabrik Reutter, die gerade in Zuffenhausen ein neues Werk eingerichtet hatte. Das Gebäude, in dem diese Volkswagen-Prototypen entstanden sind, existiert noch heute und ist Bestandteil des Porsche-Werk 2.

Am 26. Mai 1938, dem Himmelfahrtstag, hatten die drei V303-Wagen ihren großen Auftritt: In einer

Die Reutter & Co. GmbH baute die Karosserien der Vorserie V303
Reutter & Co GmbH built the bodies for the V303 pre-production cars

Another worthwhile improvement to the V303's body was that the doors were now hinged to the A-posts. Chief stylist Erwin Komenda, who was in charge of Volkswagen body development, had already suggested this change, but it was not adopted until Ferdinand Porsche, on one of his trips to the USA, noticed that almost all new cars in that country had front-hinged doors. The V303 also had bumpers, though these had been fitted experimentally to only a few of the W30 cars. To simplify production, the V303's trunk lid was extended down; this also improved the car's appearance. The bodies for this latest batch of cars were not built by Daimler-Benz but by Reutter, the Stuttgart-based company, which had just built and equipped a second factory in Zuffenhausen. The building in which these Volkswagen prototypes were constructed is still in existence as part of Porsche's #2 plant.

On May 26, 1938, Ascension Day and therefore a public holiday in Germany, the three V303 cars made their first grand public appearance. An opulent propaganda show had been organized for 70,000 invited guests to celebrate laying the foundation stone of the new

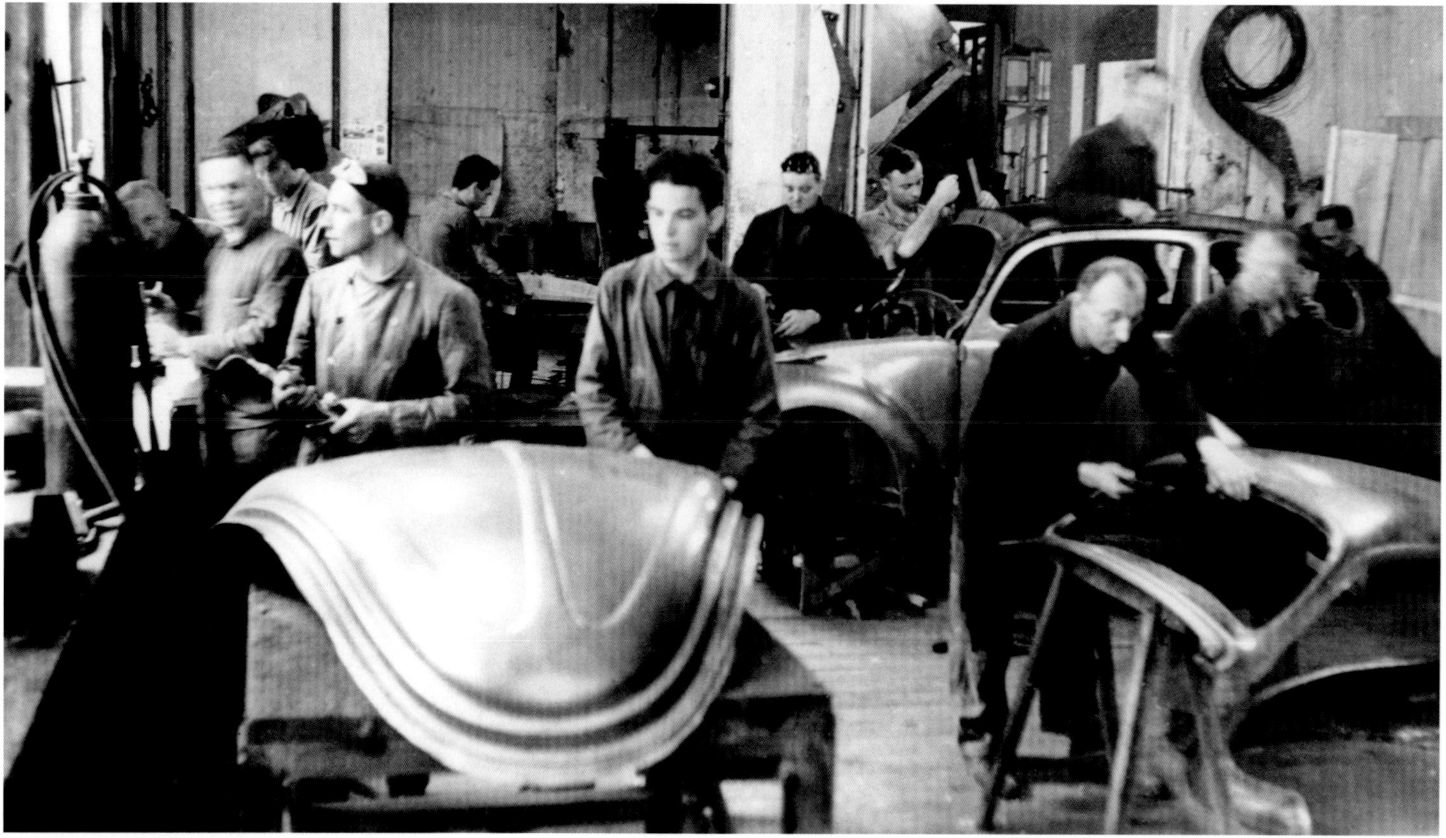

IIIA-42801
IVB-109515
NSU

Linke Seite: Die V303-Limousine 1938 im Hof des Porsche-Werk 1
Left page: The sedan (one of the three V303 cars) in the yard of #1 plant, 1938

Der Vierzylinder-Viertakt-Boxermotor des V303
The V303's flat four four-stroke engine

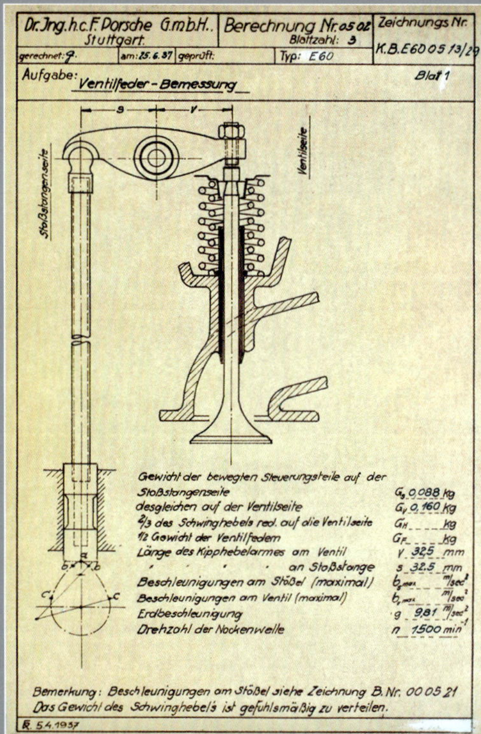

Berechnung der Feder der Ventilsteuerung
Calculating the valve spring rate

aufwendigen Propagandashow wurde vor 70.000 geladenen Gästen der Grundstein zum Volkswagenwerk gelegt, wobei auch die neuen Wagen der Öffentlichkeit präsentiert wurden. Anlässlich der feierlichen Ansprache wurde der Volkswagen von Adolf Hitler ohne jede Vorankündigung mit einem neuen Namen bedacht: Sehr zum Missfallen von Ferdinand Porsche sollte das Fahrzeug fortan »KdF-Wagen« heißen. Hierdurch sollte die ideelle Verbindung zu anderen Projekten der »Deutschen Arbeitsfront« gestärkt und für die nationalsozialistische Freizeitorganisation »Kraft durch Freude« geworben werden.

Der Ort des Geschehens lag nahe der norddeutschen Gemeinde Fallersleben, die aufgrund ihrer verkehrsgünstigen Lage am Mittellandkanal und der Eisenbahnstrecke Ruhrgebiet-Berlin ausgewählt wurde. Am nördlichen Ufer des Mittellandkanals entstand eine 1,5 Kilometer lange Werksfront mit Verwaltungsgebäude und Fertigungshallen; daran anliegend auf der anderen Uferseite eine Wohnstadt, die am 1. Juli 1938 als »Stadt des KdF-Wagens bei Fallersleben« gegründet wurde.

production plant, the “Volkswagenwerk”. Adolf Hitler addressed the massed spectators and announced out of the blue that the Volkswagen was to have a new name: “KdF-Wagen”. Ferdinand Porsche greatly disapproved of this politically motivated move, the aim of which was to associate the project with others undertaken by the German Workers’ Front and publicize the National Socialists’ recreational organization “Strength through Joy” (hence the German initials “KdF”).

The new production plant’s site was near the North German community of Fallersleben, and had been chosen because of its favorable infrastructure, with the Mittelland Canal and the railroad from the Ruhr industrial area to Berlin close by. The frontage of the new administrative and production buildings on the north bank of the canal was one and a half kilometers long. On the other side of the canal, work started on a new town to be known by the rather unlikely name of “Stadt des KdF-Wagens bei Fallersleben” (KdF Car Town near Fallersleben”).

Das Interieur des V303 war nahezu identisch mit dem Armaturenbrett des »VW-Käfers« aus der Nachkriegsproduktion
The V303’s interior, now almost identical for the first time with the post-war “VW-Beetle’s” dashboard

Die Cabriolet-Version des V303, 1938
The convertible version of the V303, 1938

Parallel zu den Bauarbeiten am Volkswagenwerk näherte sich der »KdF-Wagen« seiner endgültigen Version. In der ersten Juliwoche 1938 nahm Ferry Porsche mit einer V303-Limousine sowie einem Cabriolet an einer Alpen-Vergleichsfahrt teil, wobei sich die Porsche-Konstruktion den Vergleichsfahrzeugen Opel Kadett, Adler Trumpf Junior oder DKW Meisterklasse überlegen zeigte. Besonders die enorme Steigfähigkeit des V303 fiel positiv auf. An den Steigungen des Großglockners, bei denen die Motoren der durchweg mit Wasserkühlung versehenen Vergleichswagen zu kochen begannen, fuhr der luftgekühlte VW ohne Probleme bis zum Gipfel weiter.

Das Interieur des V303-Cabriolets mit Zusatzinstrumenten für Testfahrten
Inside the V303 convertible, with additional instruments for testing purposes

Rechte Seite: Die »Rolldachlimousine« des V303, 1938
Right page: The V303 as a sedan with roll-back roof, 1938

As construction work on the new production facility progressed, the final version of the "KdF Car" also took shape. In the first week of July 1938 Ferry Porsche took a V303 sedan and a convertible to the Alps for a comparative road test. The Porsche design was pitted against the Opel Kadett, Adler Trumpf Junior and DKW Meisterklasse, and proved superior to them all. The V303's immense hill-climbing capabilities were especially obvious. When climbing the Grossglockner Pass, the rival cars' water-cooled engines tended to overheat, whereas the air-cooled Volkswagen reached the summit of the pass without the slightest difficulty.

IIIA - 42802

IIIA-42803

Linke Seite: Ferdinand Porsche (links) während einer V303-Alpentestfahrt im September 1938
Left page: Ferdinand Porsche (left) during testing in the Alps, September 1938

Ferry Porsche am Steuer des V303-Cabriolet
Ferry Porsche in the V303 convertible

IIA - 43006

SERIENREIF: DER VW38

Mit der Version VW38 hatte der Volkswagen Mitte des Jahres 1938 seine endgültige Gestalt gefunden. Als Kleinserie entstanden im neuen Porsche-Werk 1 in Zuffenhausen weitere 30 Fahrzeuge, deren Karosserien von Reutter zugeliefert wurden. Auch international fand der VW nun Aufmerksamkeit; selbst die »New York Times« berichtete über das deutsche Volksautomobil. In einem Artikel bezeichnete sie am 3. Juli 1938 den Wagen ironisch als »Käfer« – das erste Mal, dass dieser Begriff öffentlich verwendet wurde.
Ab der zweiten Jahreshälfte 1938 wurde das Marketing für den Volkswagen intensiviert. Medienwirksam rollte eine Kolonne von VW38 durch das Deutsche Reich um Werbung für den »KDF-Wagen« und sein neuartiges Sparsystem zu machen. Interessenten konnten ab dem 1. August 1938 monatlich fünf Mark auf ein Sparkonto einzahlen, bis der Gesamtbetrag von 990 Reichsmark erreicht war. Bis Ende 1939 wurden rund 170.000 Sparanträge gestellt, von denen jedoch kriegsbedingt kein einziger bis zur Auszahlung in Form eines Autos kam.

Die Vorserie VW38 bestand aus 30 Fahrzeugen. Hier eine Limousine auf der Autobahn nahe Stuttgart (1938)
The VW38 pre-production run consisted of thirty cars; this sedan is seen on the 'autobahn' near Stuttgart in 1938

READY FOR PRODUCTION: THE VW38

By midway through 1938 the Volkswagen had reached its final form. This was the VW38, of which a batch of 30 was built at the new Porsche #1 plant in Zuffenhausen, with bodies supplied by Reutter. The Volkswagen now began to attract international attention: on July 3, 1938 the New York Times devoted an article to the new German people's car and described it ironically as a "Beetle" – the first public use of this term. In the second half of 1938, marketing for the Volkswagen was intensified. Effective publicity was obtained by dispatching a cavalcade of VW38 cars through Germany to draw public attention to the "KdF Wagen" and the novel saving scheme introduced for intending buyers. Starting on August 1, 1938, savers were invited to deposit five Reichsmarks monthly on a savings account until the full purchase price of 990 Reichsmarks was reached. By the end of 1939, about 170,000 people had begun to save for a Volkswagen in this way, but after the war broke out none them ever received the cars they had paid for.

Die Evolution des Volkswagens:
The evolution of the Volkswagen:

Die Heckansichten zweier V3
Rear views of two V3 cars

Die Heckansicht des W30
Rear view of the W30

Die Heckansicht des VW38
Rear view of the VW38

Linke Seite: »Rolldach-limousine« aus der Vorserie VW38 (1938)
Left page: Pre-production VW38 sedan with roll-back roof, 1938

Das spartanische Interieur des VW38 (1938)
The sparsely equipped interior of the VW38 in 1938

Die Drehstabfederung des VW38. Porsche hat diese Federung 1931 patentieren lassen
The VW38's torsion-bar suspension was a Porsche invention

Rechte Seite: Ferdinand Porsche (links) und Karl Rabe vor einer Zeichnung des VW38
Right page: Ferdinand Porsche (left) and Karl Rabe in front of a drawing of the VW38

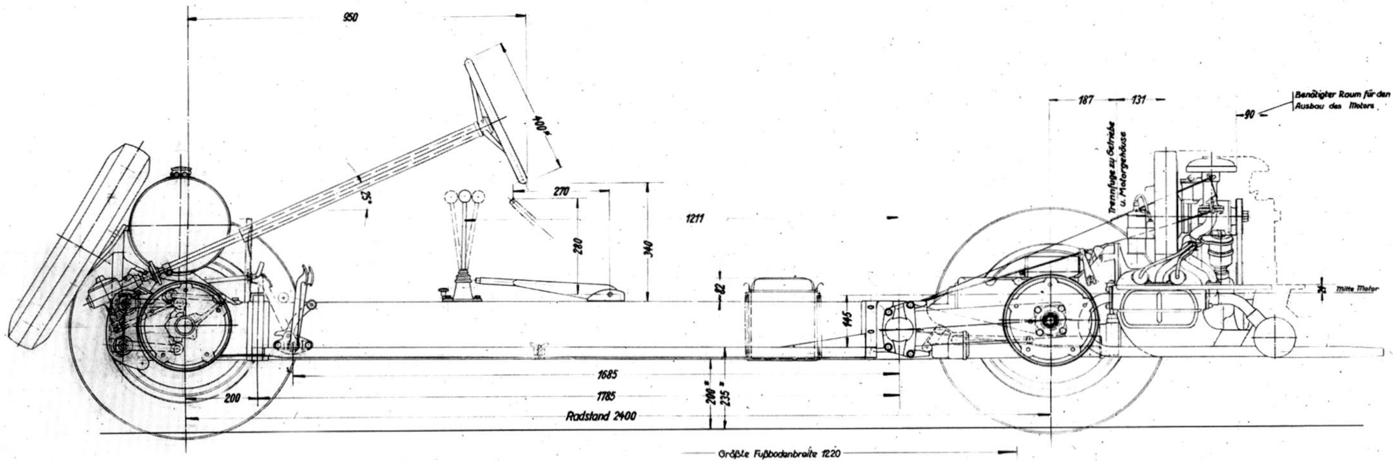

950
400
52
270
280
340
1211
82
146
187
131
90
Benötigter Raum für den Ausbau des Motors
Trennfuge zu Getriebe u. Motorgehäuse
Mitte Motor
1685
1785
200
200
235
Radstand 2400
Größte Fußbodenbreite 1220

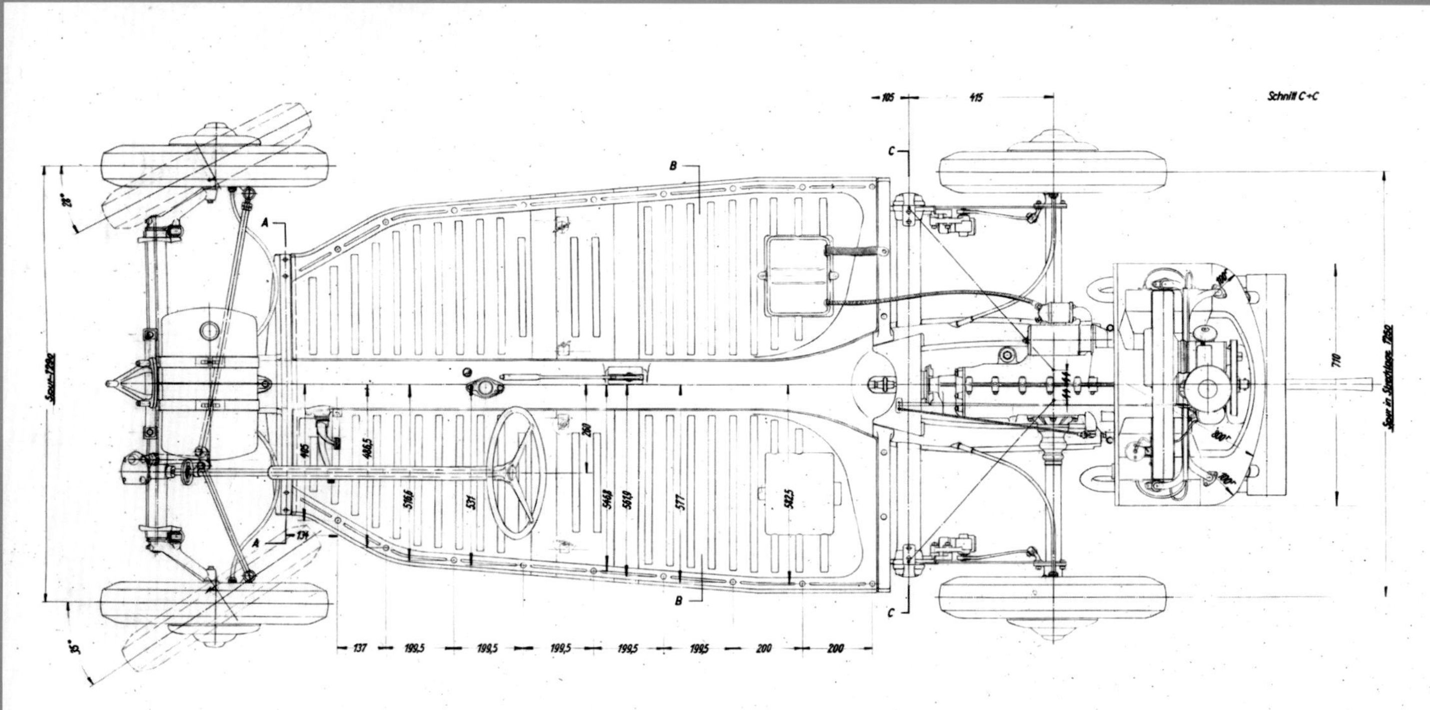

Zeichnung des Fahrgestells und Motors des VW38 vom Februar 1938

Drawing of the VW38 chassis and engine, February 1938

Wartungsarbeiten am VW38
Maintenance work at the VW38

Links: Ferry Porsche an einem Motorenprüfstand in Zuffenhausen mit dem Vierzylinder-Viertakt-Boxermotor
Left: Ferry Porsche at an engine test rig in Zuffenhausen with a flat-four four-stroke engine under test

Rechts: Motor des VW38
Right: The VW38's engine

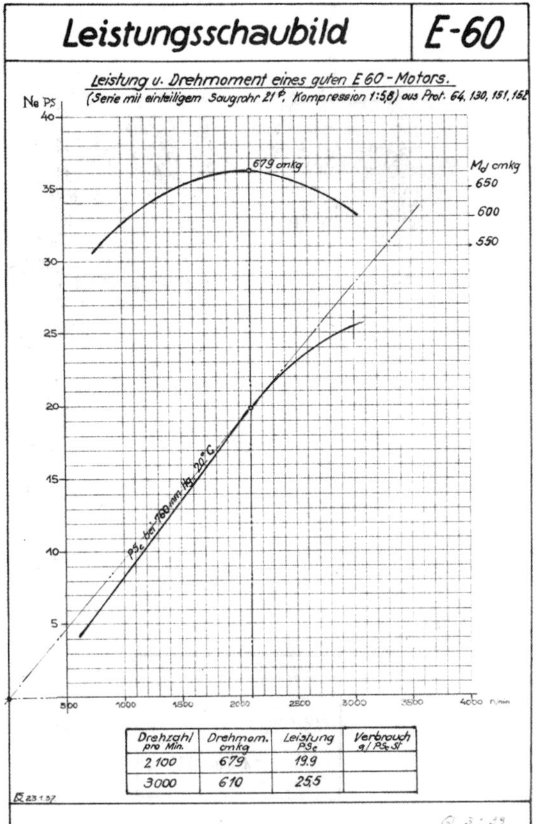
Leistungsschaubild
E-60
Leistung u. Drehmoment eines guten E 60-Motors.
(Serie mit einteiligem Saugrohr 21 ⌀, Kompression 1:5,8) aus Prot. 64, 130, 151, 152
Ne PS
40
35
30
25
20
15
10
5
679 cmkg
Md cmkg
650
600
550
PSe bei 760 mm Hg 20°C
500
1000
1500
2000
2500
3000
3500
4000
Drehzahl pro Min.
Drehmom. cmkg
Leistung PSe
Verbrauch g/PSe St
2 100
679
19.9
3000
610
25,5

D
III A
43001

Linke Seite: Leistungsdiagramm eines »guten« E60 Motors von 1938
Power-output graph of a "good" E60 engine from 1938

Alpen-Testfahrt im Juli 1938. Vorne der erste VW38, dahinter links Ferry Porsche
Test runs in the Alps, July 1938. The first VW38 is at the front, with Ferry Porsche behind it at the left

Demonstration de[redacted]de-
gängigkeit des[redacted]
Demonstr[redacted] the
VW38[redacted]oad capability

Linke Seite: Fünf Personen sollten im Volkswagen komfortabel Platz finden
Left page: The aim was for five people to travel in comfort in the Volkswagen

VW38 auf einer Ausstellung in Bietigheim 1938
The VW38 at an exhibition in Bietigheim, 1938

Diesen KdF-Prospekt erhielten Sparer im Jahr 1938

This leaflet, issued in 1938, was supplied to those who joined the savings scheme

»Motor Schau«-Ausgabe anlässlich der Automobil-Ausstellung Berlin 1939

'Motorschau' edition for the German Motor Show held in Berlin, 1939

Der erste Verkaufsprospekt des KdF-Wagen aus dem Jahr 1938
The first KdF car sales leaflet, 1938

Ausstellungs Wagen 229
IIIA-43027
Ausstellungs Wagen 228
IIIA-43026

Mehrere VW38 auf der Autobahn-Rekordstrecke bei Dessau. Ziel der Reise war die Automobil-Ausstellung in Berlin

Several VW38s on the section of the 'autobahn' near Dessau used for speed record attempts; the cars are on their way to the Berlin Motor Show

IIIA 43036

VW39 mit der Abdeckung der Frontscheinwerfer zum »Fliegerschutz«
A VW39 with the headlamps shielded as a precaution against air raids

Die Zuffenhausener Fertigungshalle 1941
The production hall in Zuffenhausen, 1941

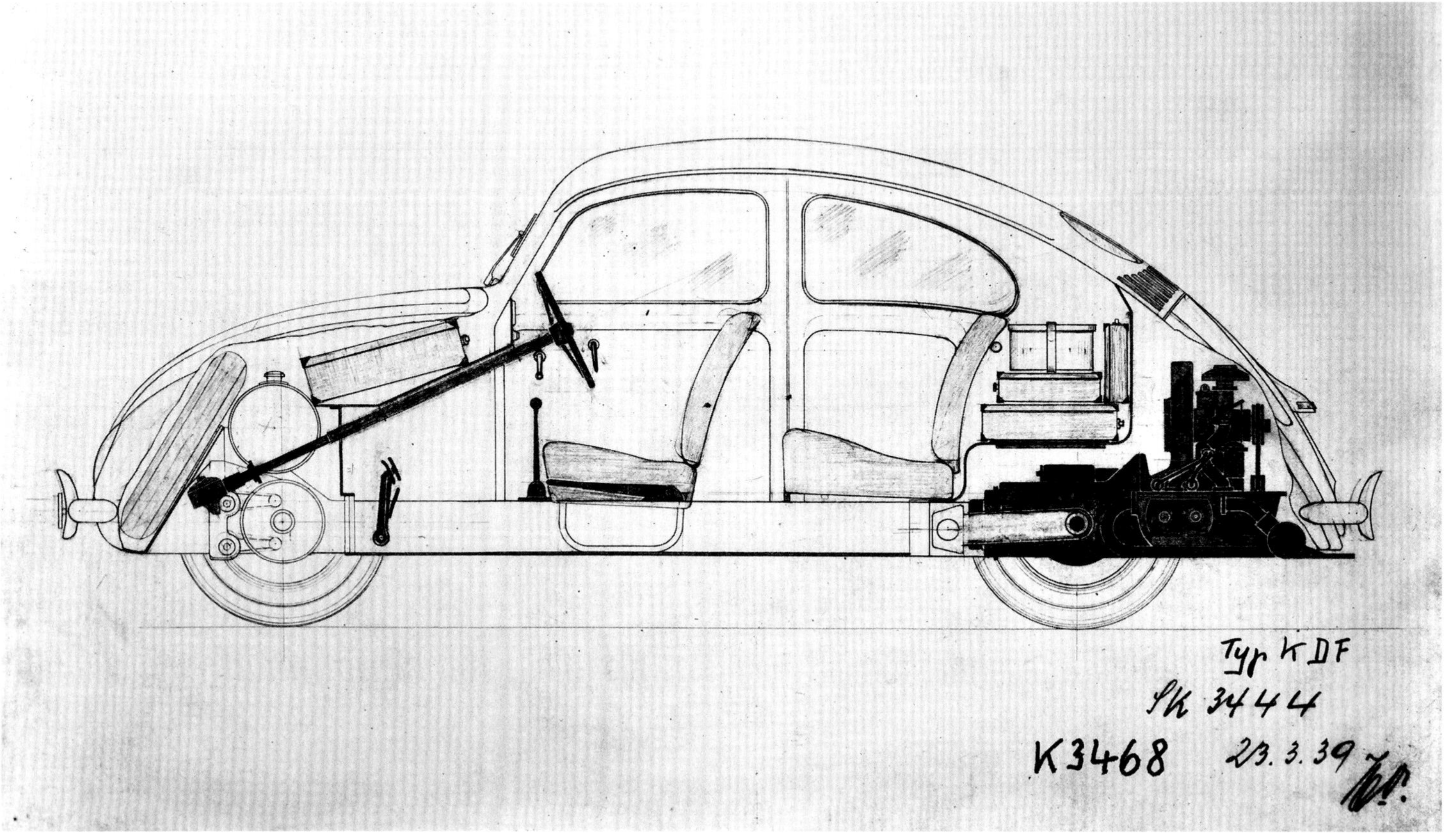
Typ KDF
Sk 3444
K3468
23. 3. 39

Auf den VW38 folgte 1938/39 mit dem VW39 eine weitere Reihe von Versuchswagen, deren Stückzahl ebenfalls in der Größenordnung von 30 Fahrzeugen lag. Die Fahrzeuge dieser Serie sind für die unterschiedlichsten Zwecke eingesetzt worden. Die »Dr. Ing. h.c. F. Porsche KG« nutzte die Wagen immer wieder zu Testfahrten, vornehmlich in den Alpen. Es wurden verschiedene Versuchseinbauten, meistens Detailänderungen am Motor, vorgenommen. Daneben wurden einige Fahrzeuge umgebaut, so dass auch Prototypen der späteren »Kübelwagen« aus dem umgebauten Typ 60 entstanden. Ebenso wurden immer wieder verschiedene Motoren mit unterschiedlichen Fahrgestellen kombiniert, weshalb schon um 1939/40 viele VW-Prototypen keine »originalen« VW38 oder VW39 mit »passenden« Fahrgestell-, Fertigungs- und Motornummern mehr waren. 1939 war die Entwicklung des Volkswagens für Porsche im Wesentlichen abgeschlossen. Als Anerkennung seiner Verdienste um den Volkswagen erhielt Ferdinand Porsche am 6. September 1938 zusammen mit Willy Messerschmitt, Ernst Heinkel und Fritz Todt den Deutschen Nationalpreis.

Die Schnittzeichnung des VW39 demonstriert den »geräumigen« Gepäckraum
A sectioned drawing of the VW39, emphasizing the trunk space and other features

Ferdinand Porsche 1940 mit seinem Volkswagen am Wörthersee
Ferdinand Porsche with his 'Volkswagen' in 1940 at his villa on Wörthersee Lake

In 1938/39 the VW38 was followed by yet another batch of prototypes, the VW 39 cars – again about 30 in all. These cars were used for a wide variety of purposes. The Dr. Ing. h.c. F. Porsche KG company regularly undertook test runs, mostly to the Alps, with them. Various experimental modifications were made, usually detail changes to the engine, but some of the vehicles were more extensively modified, for example as prototypes for the later military reconnaissance vehicles developed from the Type 60. Engines and chassis were frequently interchanged, so that by 1939/40 many of the Volkswagen prototypes were no longer "pure" VW38 or VW39 versions with matching chassis, production and engine numbers.

By 1939, Porsche had largely completed its development work. On September 6, 1938 Ferdinand Porsche was awarded the German National Prize for his contribution to the "People's Car". Other recipients for services in various areas included Willy Messerschmitt, Ernst Heinkel and Fritz Todt.

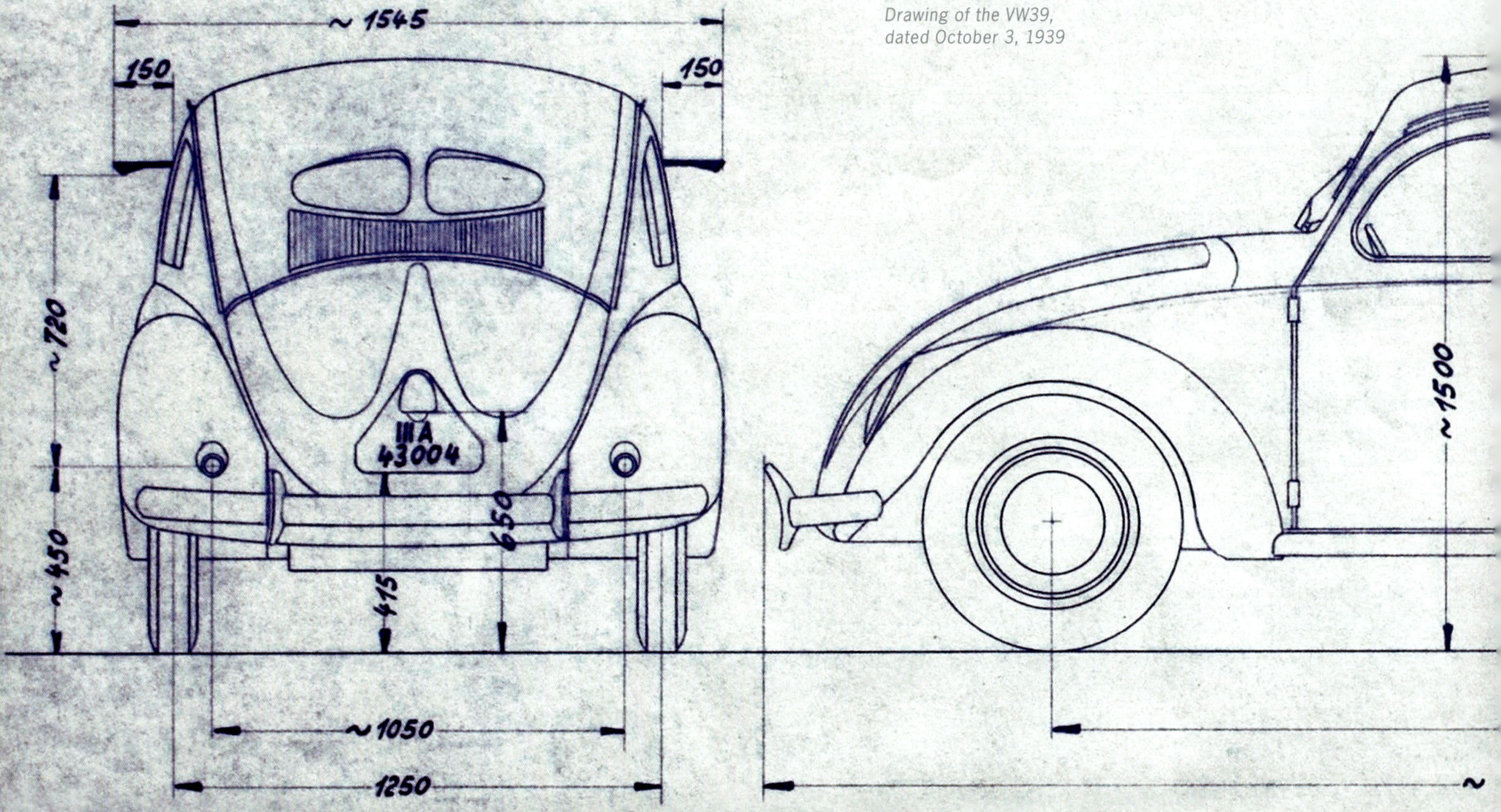

Zeichnung des VW39
vom 3.10.1939
Drawing of the VW39,
dated October 3, 1939

IIIA-43004
610
1004
1290
~1540

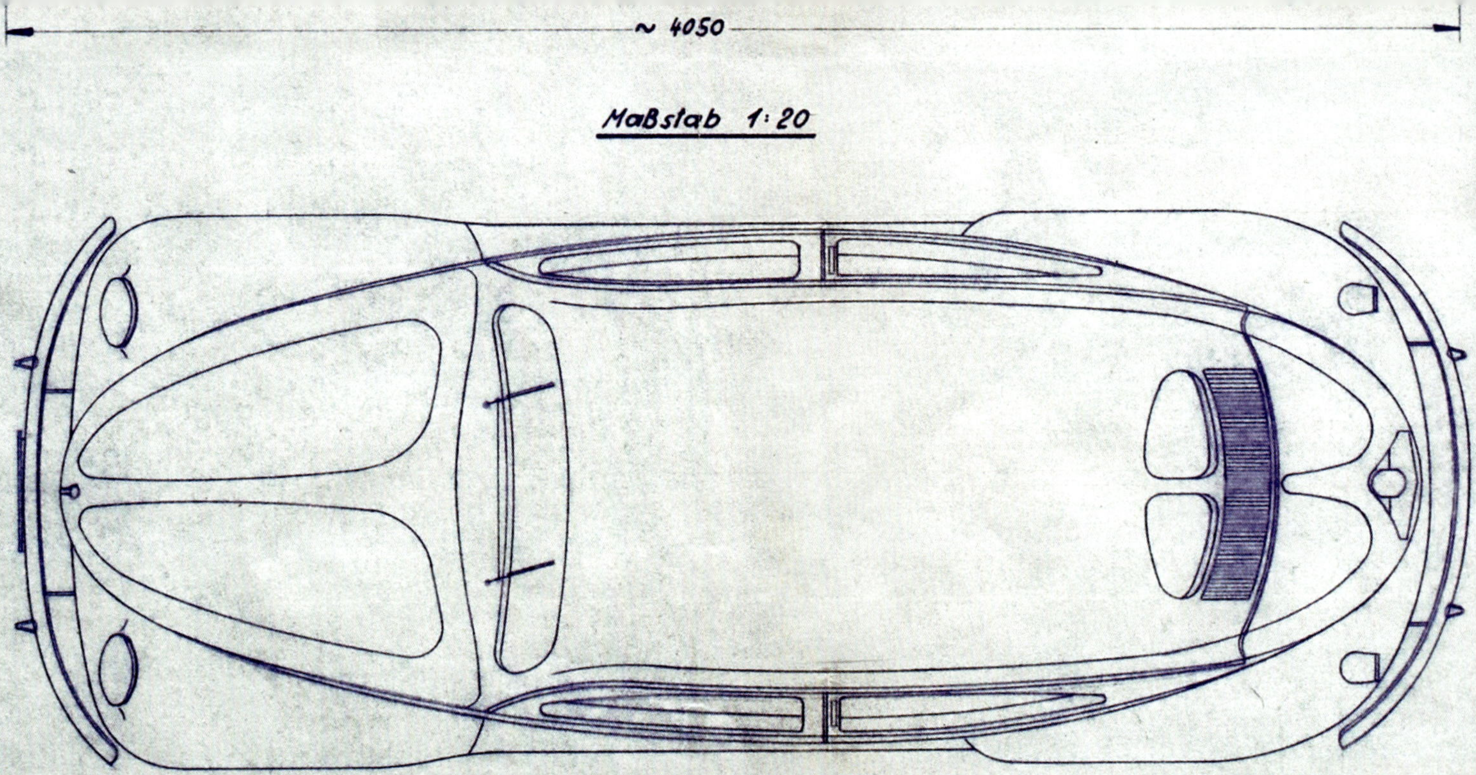
~ 4050
Maßstab 1:20

Ferry Porsches VW-Cabriolet auf der Autobahn bei Zuffenhausen

Ferry Porsche's VW convertible on the Autobahn near Zuffenhausen

WAGENWERK HERGEST
KDF-WAGEN
1 9 4 1
Hauptgeschäftsführer Dr. Lafferentz

Beginn der Serienproduktion des Volkswagens 1941 in Fallersleben, dem heutigen Wolfsburg

Series production of the Volkswagen started in 1941 in Fallersleben (name changed to Wolfsburg in 1945)

DER »BERLIN-ROM-WAGEN« TYP 64

Auf der Basis des Volkswagen Typ 60 beauftragte das Volkswagenwerk bei Porsche einen neuen VW-Rennwagentyp, der als Werbemaßnahme für den »KdF-Wagen« beim Langstreckenrennen Berlin-Rom an den Start gehen sollte. Unter der internen Bezeichnung Typ 64 beziehungsweise Typ 60K10 entwickelten und bauten die Porsche-Ingenieure im Frühjahr 1939 drei Rennsport-Coupés für die im September des Jahres geplante »Non-Stop Geschwindigkeitsprüfung«.

Da ein großer Teil des über 1.500 Kilometer dauernden Rennens auf den neuen deutschen Autobahnen stattfinden sollte, kam der Aerodynamik des Fahrzeugs besondere Aufmerksamkeit zu. Mit einer windschnittigen Stromlinienkarosserie aus Aluminium, verkleideten Radkästen sowie einem modifizierten VW-Boxermotor mit zunächst 32 PS erreichte der rund 850 Kilogramm leichte »Berlin-Rom-Wagen« eine Spitzengeschwindigkeit von bis zu 145 Stundenkilometern.

Der »Berlin-Rom-Wagen« 1939 vor der Porsche-Villa
The "Berlin-Rome" car in front of the Porsche villa, 1939

THE TYPE 64 "BERLIN-ROME" CAR

The Volkswagenwerk commissioned Porsche to design a new competition car based on the Volkswagen Type 60. It was to be entered for the Berlin-Rome long-distance race as a publicity vehicle for the "KdF Wagen". Three racing coupes, known internally as the Type 64 or Type 60K10, were developed and built by Porsche's engineers in the spring of 1939 for the planned "non-stop speed test", which was due to take place in September.

Most of the route, which was more than 1,500 kilometers long in all, consisted of German high-speed "autobahns", and the car therefore had to have excellent aerodynamics. With its streamlined, wind-cheating aluminum body, covered wheel arches and modified flat-four Volkswagen engine which initially developed 33 horsepower, and aided by a weight of only some 850 kilograms, the "Berlin-Rome" car could reach a top speed of 145 kilometers an hour.

Als der Kriegsbeginn die Austragung des Rennens verhinderte, benutze das Konstruktionsbüro Porsche die drei fertig gestellten Typ 64 als schnelle Reisesportwagen, mit denen auf Fahrten von Stuttgart nach Berlin Durchschnittsgeschwindigkeiten von 130 km/h erzielt wurden. Bis zu seinem ersten Rennsporteinsatz musste der einzige nach dem Krieg verbliebende Typ 64 elf Jahre warten. Im Jahr 1950 startete der Privatrennfahrer Otto Mathé mit seinem kurz zuvor erworbenen Typ 64 bei der Österreichischen Alpenfahrt und erzielte einen Klassensieg sowie die Auszeichnung für die beste Sportwagen-Gesamtleistung.

Das formschöne Coupé gilt unter Automobilhistorikern als der Urahn aller seit 1948 entstandenen Porsche-Sportwagen. Auf der technischen Grundlage des Typ 64 »Berlin-Rom-Wagen« wurden im Frühjahr 1948 die unter der Konstruktionsnummer 356 ausgefertigen Entwürfe realisiert und ein fahrbereiter Prototyp – der Porsche 356 »Nr.1« – gebaut.

Porsche Typ 64 in Gmünd (1945)
Porsche Type 64 in Gmünd, 1945

The outbreak of the Second World War prevented the race from being held, and so the Porsche design office used the Type 64 cars that had already been built as fast roadgoing sports cars capable of covering the distance between Stuttgart and Berlin at an average speed of 130 kilometers an hour. Only one of the Type 64 cars survived the war, and had to wait eleven years before it was first used in competition. In 1950 private racing driver Otto Mathé entered it for the Austrian Alpine Rally, scored a class win and collected the award for the best overall performance by a sports car.

With its attractive coupe body, the Type 64 is regarded by car historians as the ancestor of all the Porsche sports cars that have reached the market since 1948. The same technical basis as the “Berlin-Rome” car was used for the design that appeared in the spring of 1948 as project number 356. The roadgoing prototype based on this design was the legendary Porsche 356 “Number One”.

Zeichnung des Typ 64 vom
September 1938

*Drawing of the Type 64,
September 1938*

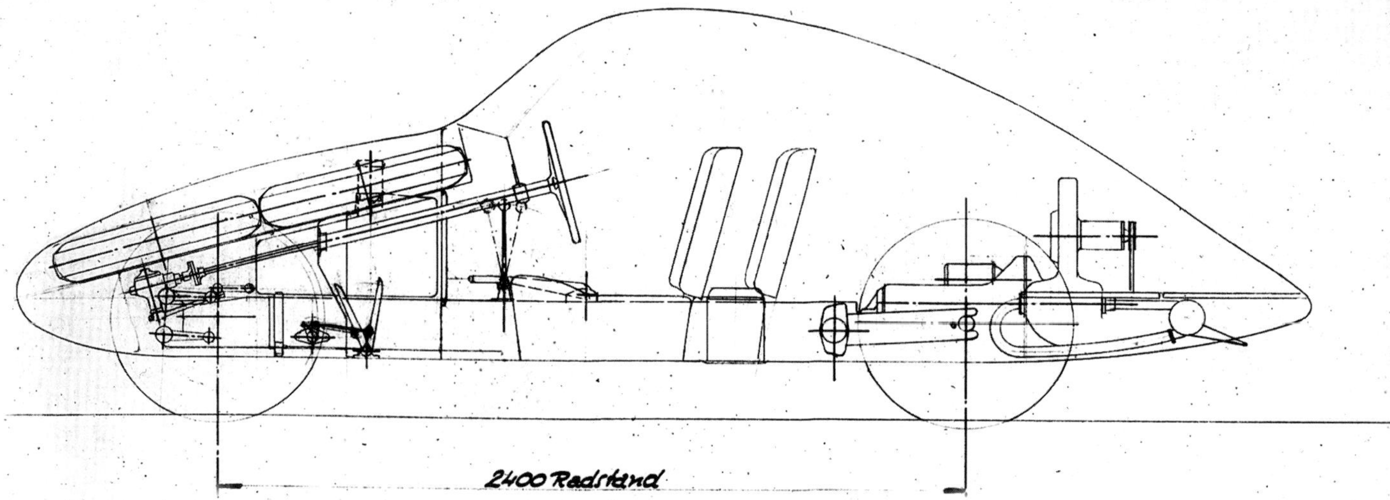

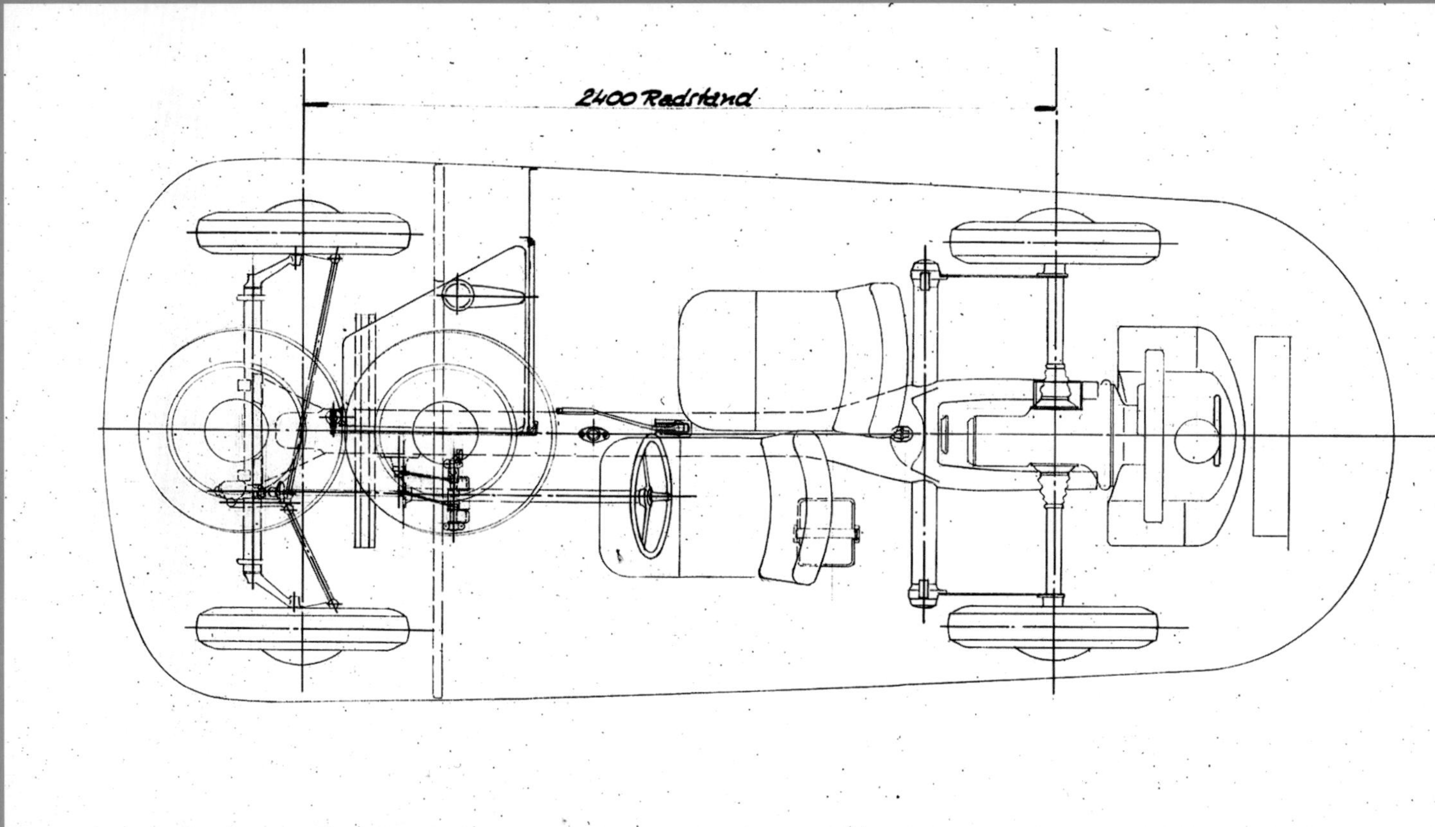
2400 Radstand

1
IIIA-0703

2
IIIA-0703

3

1 Porsche Typ 64 vor dem Werk 1 in Zuffenhausen, September 1939
Porsche Type 64 in front of the #1 plant in Zuffenhausen, September 1939

2 Der Typ 64 war aerodynamisch optimiert, um ihn für das geplante Rennen Berlin-Rom konkurrenzfähig zu machen
The Type 64 was optimized aerodynamically to make it competitive for the planned Berlin-Rome road race

3 Alle drei Typ 64 waren leicht unterschiedlich, z.B. an den Rücklichtern zu erkennen
The three Type 64 cars were not quite identical, as the rear lights show

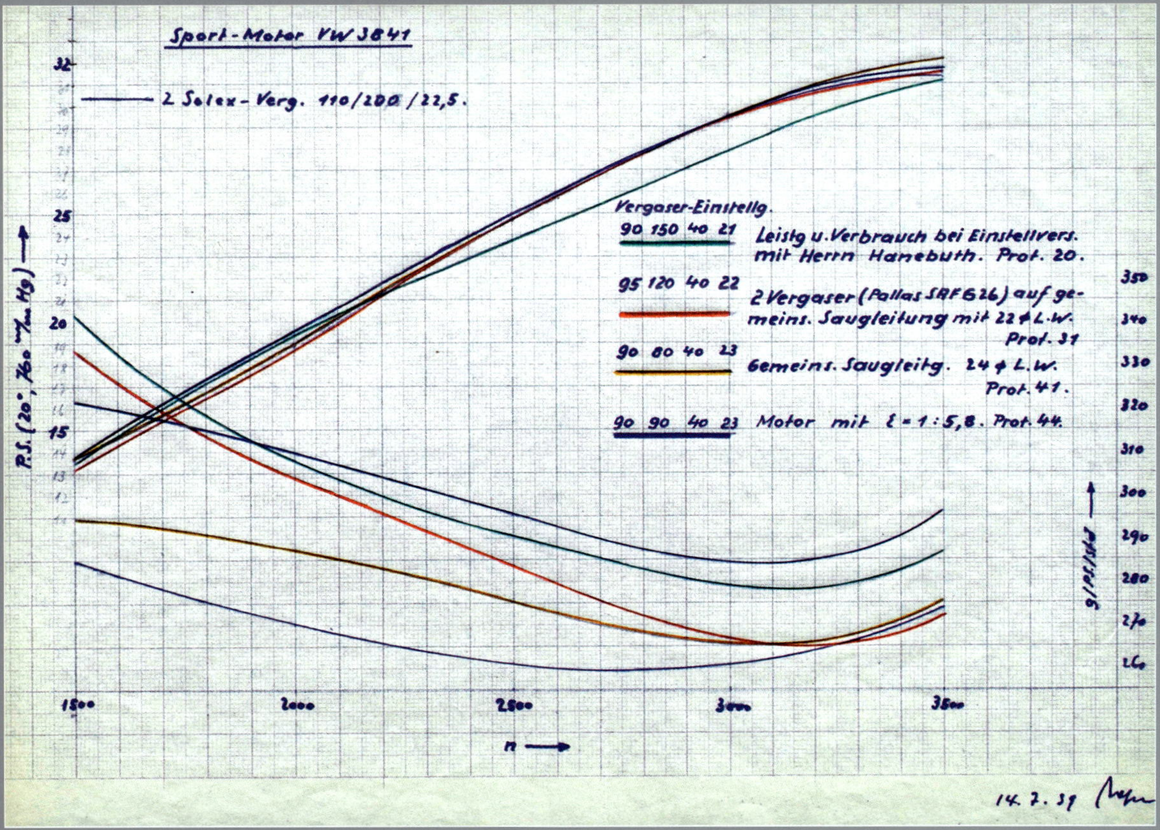

Leistungskurve des Motors für den Typ 64 mit verschiedenen Gemischaufbereitungsvarianten, Maximalleistung 32 PS
Power output graph for the Type 64 engine with various mixture supply systems; maximum power was 32 hp

Der »Berlin-Rom-Wagen« auf der Autobahn 1941
Left page: "Berlin-Rome" car on the Autobahn, 1941

Rechte Seite: Typ 64 im Hof des Porsche-Werk 1, 1941
Type 64 in the yard at Porsche #1 plant, 1941

Versuchswagen

Porsche Typ 64 in
Gmünd/Kärnten; 1949
an Otto Mathé verkauft

*Porsche Type 64 in
Gmünd (Carinthia) in 1945;
it was sold to Otto Mathé
in 1949*

Porsche Typ 64
kurz vor Kriegsende
in Gmünd/Kärnten,
1945
Porsche Type 64 in
Gmünd (Carinthia)
near the end of the
war in 1945

Otto Mathé (links) bei
der Österreichischen
Alpenfahrt 1950
Otto Mathé (left)
during the Austrian
Alpine Rally, 1950

T 2222

DER VOLKSWAGEN IM ZWEITEN WELTKRIEG

Nachdem das deutsche Heereswaffenamt die Weiterentwicklung des »KdF-Wagen« zu militärischen Zwecken angeregt hatte, wurden mehrere geländegängige Varianten des Volkswagens entwickelt. Nach dem Ausbruch des Zweiten Weltkrieges und der damit einhergehenden Umorientierung auf Rüstungsgüterproduktion wurde die Fertigung des zivilen »KdF-Wagen« vollständig auf Militärfahrzeuge umgestellt. Dies führte dazu, dass vom serienreifen Volkswagen nur wenige Fahrzeuge gebaut wurden, die ausnahmslos an Parteifunktionäre und staatliche Dienststellen gingen. Kein einziger der KdF-Sparer erhielt eines der ersehnten Fahrzeuge. Das Volkswagenwerk produzierte – auch unter dem Einsatz Tausender Zwangsarbeiter – bis zum Kriegsende 1945 ausschließlich Kriegsgerät.

Im Herbst 1938 war ein Prototyp des neuen Typ 62 »KdF-Gelände-Fahrzeug« fertig gestellt. Dessen Lastenheft umfasste unter anderem eine offene Karosserie mit geringer Bauhöhe sowie ein Gesamtgewicht von maximal 950 Kilogramm inklusive

Eine frühe Variante des Geländewagen, Typ 62
Early version of the Type 62 off-road vehicle

VOLKSWAGEN IN THE SECOND WORLD WAR

As the German army's purchasing office had proposed that work on the "KdF Wagen", as it was then called, should include vehicles for military purposes. Several off-road versions of the forthcoming "People's Car" were developed. The outbreak of war meant that industrial capacity in Germany was immediately switched over to military requirements, after which the civil-market car was supplied only to the military authorities. Only a few Volkswagen sedans were built, although the car was ready for volume production at that time. Those that did leave the assembly lines were without exception allocated to the political party for use by its officials, or to other government departments. None of the people who saved loyally month by month ever received one of the cars the has hoped for. Until hostilities ceased in 1945, the Volkswagenwerk produced military vehicles only, and employed forced laborers by the thousand to do so.

In the last months of 1938, the first prototype of the new Type 62 was completed: this was the "KdF Offroad Vehicle". Its specification called for an open

Besatzung und Ausrüstung. Nach einer erfolgreichen Fahrerprobungsphase entschied sich das Oberkommando der Wehrmacht für die Serienfertigung des inzwischen als Typ 82 »VW-Kübelwagen« bezeichneten Fahrzeugs. Anfang 1940 wurde im Volkswagenwerk mit der Produktion begonnen, wobei die Karosserien von Ambi-Budd aus Berlin zugeliefert wurden. Als das Standardgeländefahrzeug der deutschen Wehrmacht erlangte der »Kübelwagen« große Bekanntheit. Weitere Varianten waren der Typ 81 »VW-Kastenwagen« sowie der mit Allradantrieb ausgestatteten Typ 87. Zwischen 1939 und 1945 wurden im Volkswagenwerk über 50.000 Militärfahrzeuge in verschiedenen Ausführungen hergestellt.

Im Juni 1940 erhielt die Porsche KG vom Heereswaffenamt einen Entwicklungsauftrag über ein schwimmfähiges Geländefahrzeug, das auf der Grundlage des vierradangetriebenen »Kübelwagen« entwickelt werden sollte. Ein erster Prototyp dieses Typ 128 entstand im Herbst 1940 in Kooperation mit dem Heilbronner Karosseriebauer Drauz und wurde zunächst auf dem Feuerlöschteich des Porsche-Werksgeländes in Zuffenhausen erprobt. Anfang November 1940 wurden

Drei Typ 62, noch mit in die Frontpartie integrierten Ersatzrädern, neben einem Typ 92 im Werk 1 in Zuffenhausen
Three Type 62s, still with the spare wheel integrated into the front of the body, next to a Type 92 at #1 plant, Zuffenhausen

body of low overall height, and a total weight of no more than 950 kilograms including the occupants and their equipment. After successful trials, the Supreme Military Command decided that the vehicle, now known as the VW Type 82 "Kübelwagen" or Reconnaissance Vehicle, should go into series production. This began in 1940 at Volkswagen's plant, the bodies being delivered by the Ambi-Budd company in Berlin. As the German Army's standard off-road vehicle, the "Kübelwagen" was extremely well known. Other versions were the Type 81 with a van body and the Type 87, which had all-wheel drive. Between 1939 and 1945, the Volkswagen plant built more than 50,000 of these vehicles, in various versions.

In June 1940, Porsche KG was instructed by the German army weapons office to develop an amphibious version of the all-wheel-drive off-road vehicle. The first prototype of this Type 128 project was completed in the fall of 1940, in cooperation with the coachbuilder Drauz in Heilbronn. The first trials were held on the firefighting reservoir at Porsche's factory in Zuffenhausen. Early in November, three more Type 128s were dispatched on a lengthy test

Typ 82, vor dem Porsche-Werk 1 in Zuffenhausen, (1941)
Type 82 in front of the Porsche #1 plant in Zuffenhausen, 1941

Rechte Seite: Diese Typ 82 aus der Vorserie wurden noch im Zuffenhausener Werk 1 gebaut
Right page: This pre-production Type 82 was built at the #1 plant in Zuffenhausen

HALT

134

drei weitere Typ 128 auf einer großen Erprobungsfahrt getestet. Die Weiterentwicklung der Typen 128 »VW-Schwimmwagen« und 138 »VW-Geländewagen« führte zur Konzeption des Typ 166 »Kradschützenwagen«. Mit verkürztem Radstand und nur 900 Kilogramm schwer, war der Typ 166 außerordentlich geländegängig und sollte als Alternative zu Motorradgespannen dienen. Insgesamt wurden von 1941 bis 1944 mehr als 14.000 »VW-Schwimmwagen« hergestellt. Auch einen allradgetriebenen »Käfer« gab es im Krieg: Mit einem Fahrgestell ähnlich des Schwimmwagens Typ 128 mit langem Radstand (2400 mm) entstand 1944 der Typ 287, damals als »Kommandeurwagen« bezeichnet und in nur ganz wenigen Exemplaren gebaut.

Neben Gesamtfahrzeugen waren es zudem Detaillösungen wie ein Voll-Synchron-Getriebe, ein Dieselmotor oder eine benzin-elektrische Heizung, mit denen die Porsche KG während des Krieges im Zusammenhang mit der VW-Entwicklung beauftragt wurde. Für den Typ 82 »VW-Kübelwagen« entstand 1943 unter der Typnummer 155 eine Schneekettenlaufeinrichtung, die innerhalb kurzer Zeit montiert werden konnte. Eine Serienfertigung der Schneeraupe fand aufgrund der

Typ 87, Allradversion des Kübelwagens, 1941
Type 87, the all-wheel-drive version of the reconnaissance vehicle, 1941

run. Ongoing development work on the VW amphibian (Type 128) and VW off-road vehicle (Type 138) led to a new concept, the Type 166 "Kradschützenwagen" or light personnel carrier. With a shorter wheelbase, and weighing only 900 kilograms, the Type 166 performed extremely well off the beaten track, and was intended as an alternative to the motorcycle with sidecar. More than 14,000 VW amphibians were built between 1941 and 1944. There were even Beetle-style cars with all-wheel drive during the war years: the Type 287 was developed in 1944 with a chassis similar to the Type 128 amphibian and a long wheelbase (2400 mm), but very few of these "Commander Cars" were built.

As well as complete vehicles, Porsche KG was commissioned during the war years to develop various design details in conjunction with VW development work, for instance an all-synchromesh gearbox, a diesel engine and a combined fuel-burning and electric heater. Snow tracks that could be installed quickly were designed for the Type 82 reconnaissance vehicle in 1943, and given type number 155, but a shortage of material in wartime conditions prevented this version from entering production. In August 1943,

kriegsbedingten Material- und Rohstoffknappheit jedoch nicht statt. Auf dem Rangiergleis des Stuttgarter Hauptbahnhofs wurden im August 1943 Probefahrten mit dem Schienenfahrzeug Typ 157 durchgeführt. Die Basis dieses Fahrzeugs bildete der »VW-Kübelwagen« vom Typ 82 beziehungsweise Typ 87. Damit die Versuchswagen auch auf Eisenbahnschienen fahren konnten, wurden an den Radinnenseiten Schienenleiträder montiert.

trial runs were carried out with the Type 157 railroad version, on sidings at Stuttgart's main railroad station. This vehicle was based on the Type 82 and Type 87 reconnaissance vehicles, and had special rail guide wheels mounted inside the road wheels.

Links: Zwei Varianten des Typ 128 am Max Eyth-See in Stuttgart im April 1941
Left: Two versions of the Type 128 at Max Eyth Lake, Stuttgart, in April 1941

Rechte Seite: Beim Typ 287 war das höher gelegte Allradfahrwerk (Typ 87) mit der Käferkarosserie verbunden
Right page. For the Porsche Type 287 the all-wheel-drive chassis with its extra ground clearance (Type 87) was combined with the sedan body

Rechts: Erprobungsfahrt Typ 87, dahinter ein Typ 92
Right: Test run with a Type 87. In the backround a Type 92.

IIIA-0842

Der Typ 128 war ein Schwimmwagen mit dem Radstand des Volkswagens, hier 1940 mit Ferry Porsche am Steuer
The Type 128 was an amphibian but retained the original Volkswagen's wheelbase; it is seen here in 1940 with Ferry Porsche at the wheel

Ferdinand Porsche (rechts) bei einer Versuchsfahrt eines Typ 128, 1940
Ferdinand Porsche (right) during a test run with the Type 128, 1940

Technische Zeichnung des Typ 128 vom Mai 1941

Technical drawing of the Type 128, dated May 1941

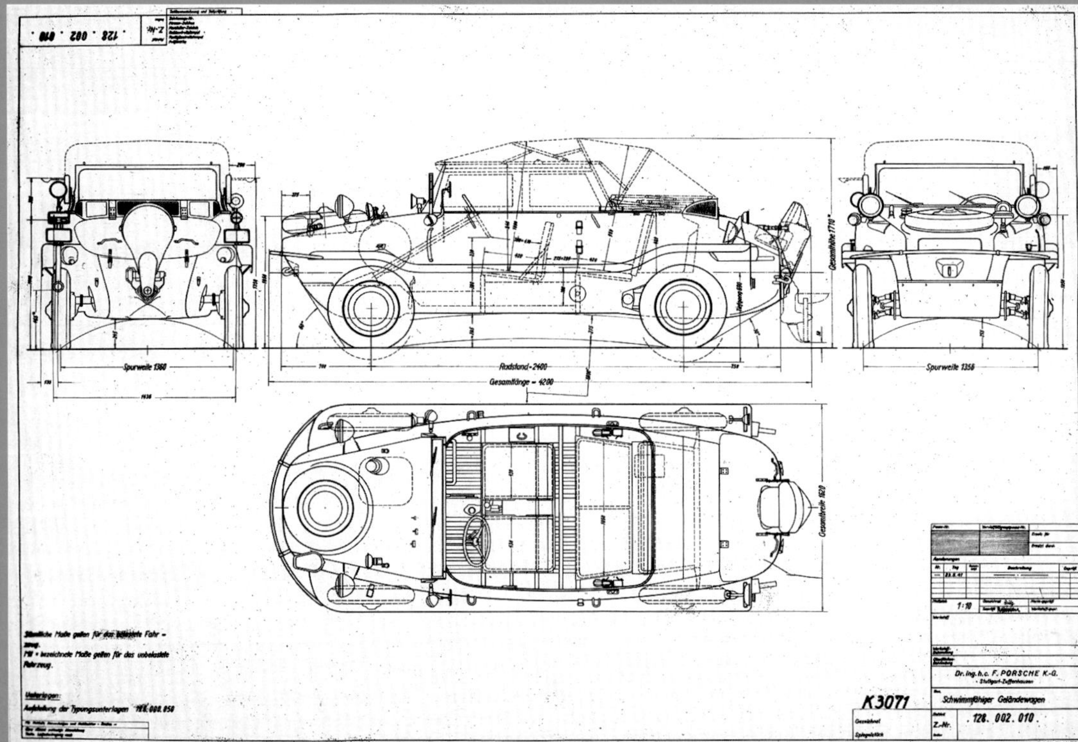

Rechte Seite: Versuch des Schwimmwagens im Juli 1941 im Max-Eyth-See in Stuttgart

Right page: Testing the amphibian at Max Eyth Lake, Stuttgart, in July 1941

Die ersten 125 Stück des Typ 166 wurden 1942 in Zuffenhausen gebaut
The first 125 Type 166 vehicles were built by Porsche in Zuffenhausen

Rechte Seite: Prüfung des Schwimmwagens für die Heeresabnahme im Mai 1942
Right page: Testing the amphibian for acceptance by the army in May 1942

IIIA-0831
IIIA-0811

Typ 166 in Zuffenhausen, um 1941

The Type 166 in Zuffenhausen, 1941

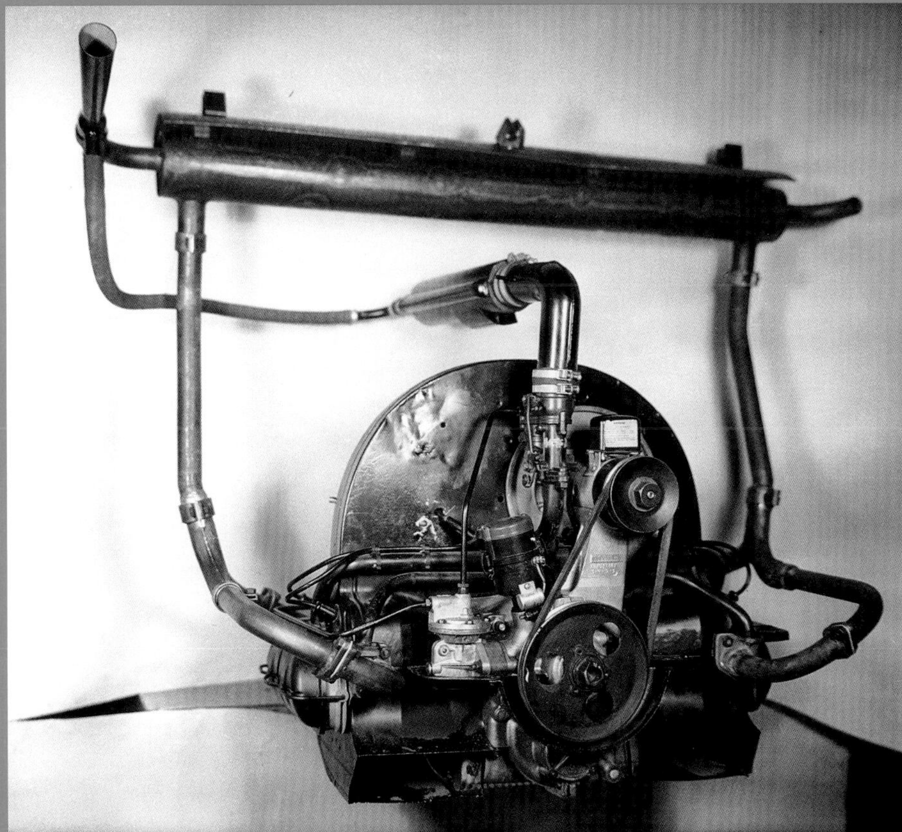

Der Motor des Typ 166 hatte eine speziell auf den Schwimmbetrieb ausgerichtete Abgasanlage
The exhaust system for the Type 166's engine was specially laid out for waterborne travel

Der Typ 166 war ein Schwimmwagen mit dem verkürzten Radstand, um die Geländetauglichkeit zu erhöhen
The Type 166 was the amphibian with a shorter wheelbase, which increased its all-terrain capability

Rechte Seite: Typ 166 mit heruntergeklapptem Wasser-antrieb (1941)
Right page: Type 166 with the screw lowered for travel on water, 1941

Linke Seite: Erprobung des Typ 166 mit Schneetrommeln am Großglockner im Juni 1942
Left page: Testing the Type 166 with drum wheels for use in snow at the Grossglockner in June 1942

Ein Schwimmwagen Typ 166 in Zell am See mit Ferdinand Porsche (links)
A Type 166 photographed in Zell am See, with Ferdinand Porsche (left)

Zur Steigerung der Geländegängigkeit wurde der Typ 155 konstruiert, ein Geländewagen mit Kettenantrieb

For greater all-terrain capability the tracked Type 155 was designed

Typ 82 mit speziellen Sandreifen

Type 82 with sand wheels

WH
816 127

VW & PORSCHE:
EINE GESCHICHTE DER ZUSAMMENARBEIT
VW & PORSCHE:
A HISTORY OF COOPERATION

K 49.801

VW & PORSCHE: EINE GESCHICHTE DER ZUSAMMENARBEIT

VW AND PORSCHE: A HISTORY OF COOPERATION

Der gemeinsame Weg von Porsche und Volkswagen wurde auch nach dem Ende des Krieges weiter fortgesetzt. Unter Verwendung von zahlreichen Bauteilen des Volkswagens entstand 1948 der erste Sportwagen mit dem Namen Porsche, der Typ 356 »Nr.1«. Auch die Serienversion des Porsche 356 basierte auf dem VW, von dem Fahrwerk, Motor und Getriebe in modifizierter Form übernommen wurden. Die Beziehungen zwischen Porsche und VW wurden am 17. September 1948 offiziell fixiert, als Ferry Porsche und Heinrich Nordhoff, Generaldirektor der Volkswagenwerke GmbH, einen zukunftsweisenden Vertrag aushandelten: Neben einer Lizenzgebühr von 5 DM pro gefertigten »VW-Käfer« verpflichtete sich Volkswagen zur Lieferung von Bauteilen für Porsche-Sportwagen und gestattete ferner die Nutzung der VW-Vertrieb-

Porsche 356/2 vor dem Typ 356 »Nr.1« auf dem Porsche-Werksgelände in Gmünd (1948)
Porsche 356/2 in front of Type 356 "Number One" on the Porsche plant site in Gmünd (1948)

Porsche and Volkswagen maintained their mutual progress in the post-war years. Many Volkswagen components were used in the first sports car to bear the Porsche name, Type 356 "Number One", which was built in 1948. The production version of the Porsche 356 was also based on the Volkswagen, from which it adopted the suspension, engine and gearbox in modified form. On September 17, 1948 the relationship between Porsche and VW was officially finalized in a forward-looking contract negotiated by Ferry Porsche and Heinrich Nordhoff, the Managing Director of Volkswagenwerke GmbH. Volkswagen undertook to pay a license fee of five Deutschmarks for every VW Beetle it built, agreed to supply parts for Porsche sports cars and also allowed Porsche to make use of the VW sales organization and after-sales

sorganisation und des Service-Netzes. Im Gegenzug sollten die Porsche-Ingenieure der Wolfsburger Entwicklungsabteilung weiterhin beratend zur Seite stehen. Zudem übernahm Porsche in Österreich die VW-Generalvertretung, die Anfang des Jahres 1949 in Salzburg eingerichtet wurde.

Das 1944 kriegsbedingt nach Gmünd in Kärnten umgesiedelte Konstruktionsbüro Porsche tüftelte bereits seit 1946 wieder an neuen VW-Entwicklungen. Als die »Porsche Konstruktionen GmbH« Ende 1949 nach Stuttgart zurückkehrte, wurde an einer ganzen Reihe von Neuentwicklungen und Studien für das Volkswagenwerk gearbeitet. Dabei führte Ferry Porsche selbst die entscheidenden Verhandlungen und war maßgeblich an der Entwicklung neuer Projekte beteiligt. Sein Ziel war es, das Grundkonzept des Volkswagens weiterzuentwickeln und funktionelle Verbesserungen auf den Gebieten Raumausnutzung, Komfort, Bedienung und Fahreigenschaften durchzuführen. Die Vielfalt der von Porsche vorgelegten Entwürfe verblüfft dabei noch heute: Unter der Bezeichnung Typ 402 entstand 1949 eine »Studie über einen VW mit verkürztem Radstand und selbsttra-

Ferry Porsche (rechts) und Heinrich Nordhoff, Generaldirektor des Volkswagenwerks (1952)
Ferry Porsche (right) and Heinrich Nordhoff, Managing Director of the Volkswagenwerk (1952)

Rechte Seite: Das Volkswagenwerk im Winter 1946/1947; Volkswagen fertig zum Gebrauch
Right page: The Volkswagen plant in the winter of 1946–1947; Volkswagens ready for use

service network. Porsche engineers were to continue to advise their colleagues in the Wolfsburg development departments, and in addition Porsche took on the VW agency for Austria; its head offices were opened in 1949 in Salzburg.

The Porsche design office, which had moved to the town of Gmünd (Carinthia) in 1944, had been working on new VW developments since 1946. When it returned to Stuttgart as “Porsche Konstruktionen GmbH” at the end of 1949, it was engaged on a series of new developments and design studies for Volkswagen. Ferry Porsche conducted the crucial negotiations himself and played a major part in new project development. His aim was to extend the basic “People’s Car” concept and render it more functional, with better use of the available space, greater comfort, easier operation and improved handling. The sheer volume of draft designs submitted by Porsche is impressive even by present-day standards: in 1949 a “Study for a VW with shorter wheelbase and load-bearing bodyshell” was submitted as Type 402. Type 405 was a plan for a Volkswagen with increased power output: for this, a 32-bhp 1.2-liter four-cylinder

Linke Seite: Ferdinand Porsche am Steuer eines Volkswagens aus Nachkriegsproduktion
Left page: Ferdinand Porsche at the wheel of a post-war Volkswagen

Ferry Porsche mit den Porsche-Typen 356 (links) und 60 (rechts)
Ferry Porsche with the Porsche Types 356 (left) and 60 (right).

K 45.444
K 49.810

Ein Volkswagen aus der Nachkriegs-Serienproduktion neben zwei 356 von der Kleinserie aus Gmünd/Kärnten

A post-war series-production Volkswagen with two 356 sports cars from the small batch built in Gmünd, Carinthia

gendem Aufbau«. Als Typ 405 wurde die Studie eines leistungsgesteigerten Volkswagens geplant, für den neben einem 1,2-Liter-Vierzylinder mit 32 PS auch ein 1,8-Liter-Sechszylindermotor mit 50 PS konstruiert wurde. Ab 1951 arbeitete Porsche im VW-Auftrag an einer Querlenker-Vorderachse, einem leiseren Motor, einer verbesserten Federung und einem neuen Hinterachsgelenk. Die Umsetzung dieser technischen Ideen erfolgte jedoch erst viele Jahre später, allein das Motorgeräusch wurde mit verschiedenen Mitteln von nun an stetig reduziert. Der Verwirklichung der Porsche-Konzepte stand vor allem eines entgegen: der große Verkaufserfolg des Volkswagens, der in Qualität und Zuverlässigkeit alle vergleichbaren Autos seiner Zeit übertraf. Bereits im August 1955 war die erste Million erreicht, so dass angesichts stetig steigender Verkaufszahlen eine Strategie der konstanten Modellpflege gewählt wurde. Der »VW-Käfer« wurde bis zu seiner Produktionseinstellung im Jahr 2003 über Jahrzehnte beständig weiterentwickelt.

Parallel zu den Verbesserungsideen für den »Käfer« beschäftigten sich die Porsche-Ingenieure mit der Erarbeitung von Fahrzeugkonzepten, die das VW-Ver-

Mit dem Typ 534 entstand 1953 eine Studie für das VW-Modellprogramm
The Type 534 was a study dating from 1953 for the VW model program

Fließheck-Variante des Typ 555 aus dem Jahr 1954
A fastback version of the Type 555 (1954)

Porsche Typ 555 mit Stufenheck (1954)
Porsche Type 555 with notchback body (1954)

engine was projected, but also a 1.8-liter six-cylinder unit developing 50 hp. From 1951 on, Porsche undertook VW commissions to design a front axle assembly with lateral control arms, a quieter engine, improved suspension and a new rear swing-axle pivot. These technical concepts were only put into effect several years later, though engine noise was steadily reduced by various methods from that time on. The main factor that prevented these Porsche proposals from being implemented was quite simply the Volkswagen's enormous sales success, based on standards of quality and reliability higher than anything that other manufacturers could offer at the time. The first million "Beetles" had been built by August 1955, and since sales figures continued to rise, it is not surprising that the basic model was retained and a strategy of regular detail updates chosen. Development work on the VW "Beetle" continued steadily until production finally ceased in 2003.

In parallel with these plans for improving the "Beetle", Porsche's engineers were also developing vehicle concepts intended to enlarge the VW sales program.

Prototypen des VW 726 als Fließheck- und Stufenheck-Variante (1958)
VW 726 prototypes with fastback and notchback body styles (1958)

kaufsprogramm erweitern sollten. Mit dem Typ 534 präsentierte Porsche dem Volkswagen-Generaldirektor Heinrich Nordhoff im Herbst 1953 eine Studie zur selbsttragenden Karosseriebauweise. Der auf dem Volkswagen basierende Prototyp wies formale Ähnlichkeit mit dem Porsche 356 auf, war mit nur 26 PS jedoch deutlich schwächer motorisiert. Eine weitere Volkswagen-Studie war der 1953/54 entwickelte Typ 555, von dem Prototypen mit Fließ- und Stufenheck gebaut wurden. Mit dem Entwicklungsziel der Sparsamkeit entstanden verschiedene Dieselmotor-Entwicklungen der fünfziger Jahre. Neben einer Diesel-Version des Käfer-Motors zählte dazu die Konstruktion des Typ 619 »Zweizylinder-Zweitaktdiesel mit Spülpumpe«, der bei 600 ccm immerhin 20 PS leistete. Da das Interesse an Dieselmotoren bei Volkswagen damals jedoch gering war, wurden diese Entwicklungen nicht weiter verfolgt.
Eine Untersuchung über das Potenzial eines Kleinwagens mit luftgekühltem Unterflurmotor entstand 1955 mit der Vorstudie Typ 672. Sein Sechszylinder-Boxermotor sollte bei 1,2 und 1,5 Liter Hubraum 43 beziehungsweise 54 PS leisten. Der Typ 675 war eine umfangreiche Neuentwicklung mit zahlreichen

Versuchsausführungen des Typ 728 im Zuffenhausener Porsche-Werk (1958)
Experimental versions of the Type 728 at Porsche's plant in Zuffenhausen (1958)

The Type 534 design study with a load-bearing bodyshell was presented to Volkswagen's Managing Director Heinrich Nordhoff in the fall of 1953. Based on the Beetle, the prototype resembled the Porsche 356 in appearance, but had a less powerful engine rated at 26 hp. Another Volkswagen design study was the Type 555, developed in 1953/54; prototypes with a fastback and a notchback body were built. With fuel economy in mind, various diesel engines were developed in the 1950s. One was a diesel version of the Beetle engine, another the Type 619 "Twin-cylinder two-stroke diesel with scavenge pump", which had an impressive output of 20 hp from a displacement of only 600 cc. At that time, however, Volkswagen exhibited very little interest in diesel engines, and so none of these projects was pursued further.
The 1955 Type 672 preliminary study investigated the potential of a small car with an air-cooled underfloor engine. This was to be a flat six with 1.2 or 1.5 liters' displacement and a power output of either 43 or 54 hp. The Type 675 was an extensive new development, completed in various forms between 1955 and 1959. A small car, it was originally planned to have a 600-cc engine; shortly after this, a flat-four version

5
4

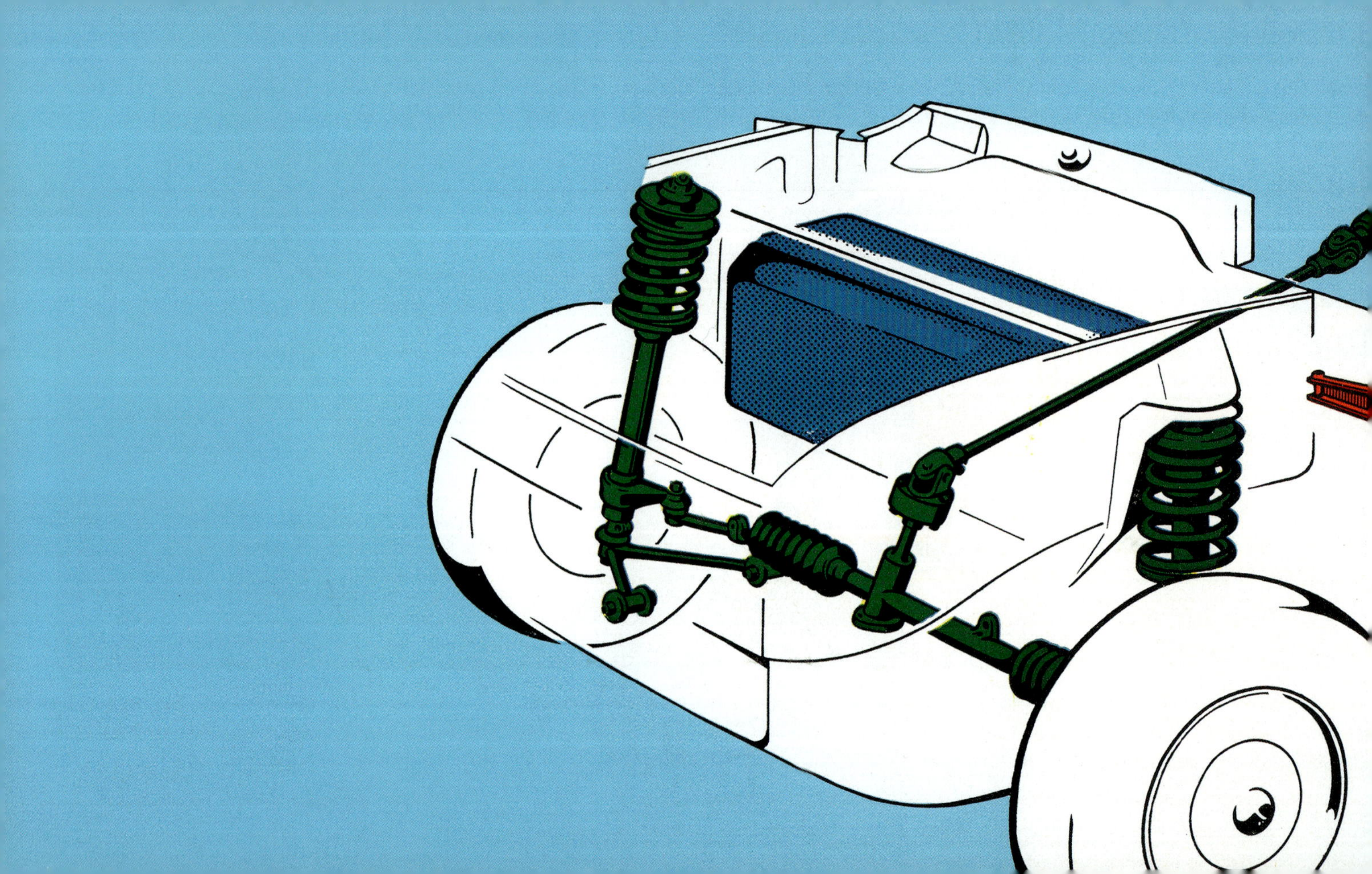

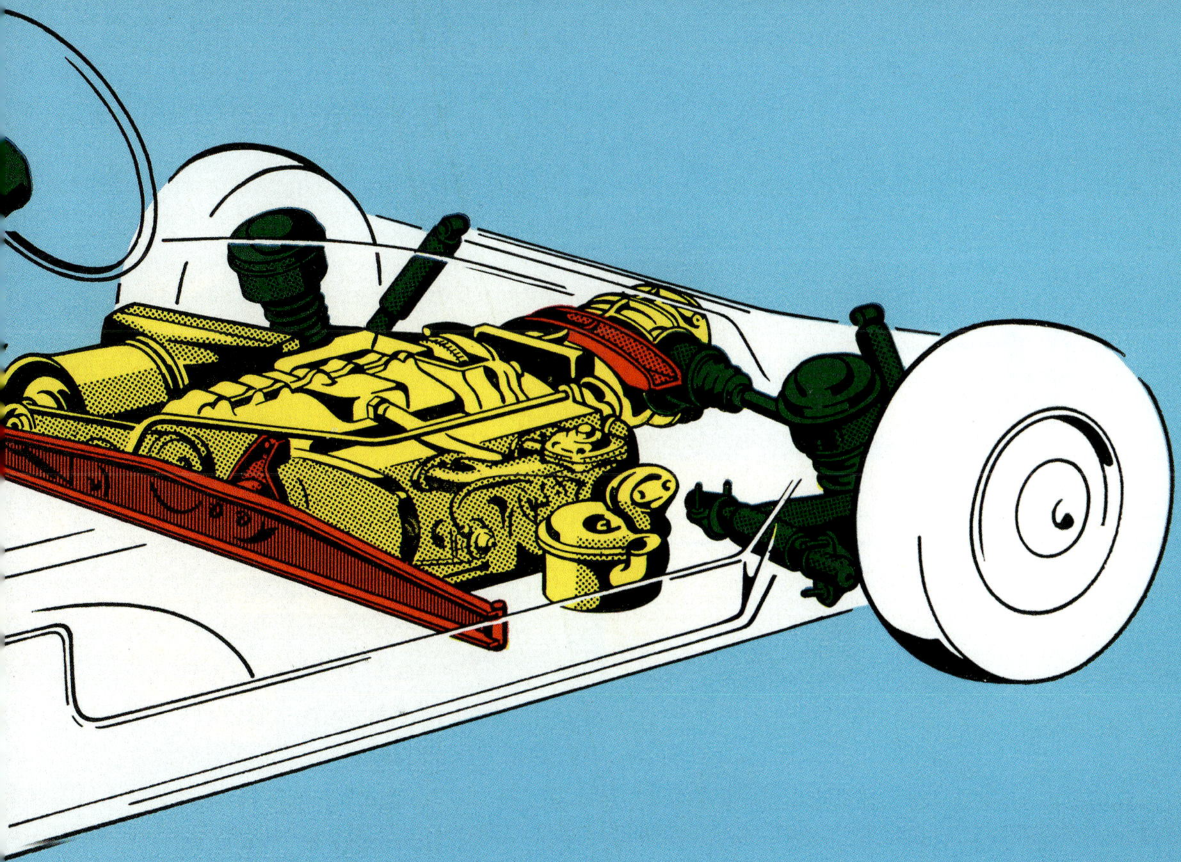

Strichzeichnung des Porsche Typ 1966, Bodengruppe und Fahrwerk (ca. 1969)
Drawing of the Porsche Type 1966's platform and suspension (about 1969)

Varianten, die zwischen 1955 und 1959 umgesetzt wurden. Als Kleinwagen war er zunächst mit einem 600 ccm-Motor projektiert, dem wenig später ein Vierzylinder-Boxer als Heck-Unterflurmotor folgte. Besonders fortschrittlich waren seine Querlenker-Vorderachse und Schräglenker-Hinterachse, die erst deutlich später im Automobilbau üblich wurden. Trotz seines Radstands von nur 2,08 Meter war der Typ 675 ein geräumiges Auto, das ein besonders gutes Preis-Leistungs-Verhältnis bieten sollte. Ein weiterer Kleinwagen-Entwicklungsauftrag aus Wolfsburg war der Typ 728, von dem zwischen 1957 und 1960 Prototypen mit verschiedenen Karosserieformen entstanden. Insbesondere die Frontpartie des selbsttragenden Aufbaus zeigte bereits die Gestalt des 1961 vorgestellten Mittelklassemodells »VW Typ 3«. Als potenzieller Nachfolger des »VW-Käfers« in Stufen- und Fließheckvariante konzipiert, kam auch dieses Projekt nicht über das Erprobungsstadium hinaus. Unter der Typnummer 1966 wurde von 1966 bis 1970 ein Fahrzeugprogramm entwickelt, das von einer 2-, 3- und 5-türigen Limousine über ein 2+2 Coupé und einem Roadster bis hin zu einem Kleinbus reichte. Sämtliche Varianten sollten sich dabei die sel-

Coupé-Variante des Porsche Typ 1966
Coupe version of the Porsche Type 1966

was sketched out with the engine under the floor at the rear. The suspension was especially advanced, with lateral control arms at the front and semi-trailing arms at the rear, a layout that was not widely used in the car industry until some years later. Although the wheelbase was only 2.08 meters long, the Type 675 had plenty of interior space and ample performance potential for the intended selling price. Another small-car design commissioned by Wolfsburg was the Type 728, of which prototypes with various body styles were built between 1957 and 1960. The front end of the load-bearing bodyshell in particular resembled the midsize “VW Type 3” model first seen in 1961, and intended as a potential successor to the Beetle. Notchback and fastback versions were built, but again the project never got beyond the initial trial stage. Type number 1966 stood for a program of vehicles developed between 1966 and 1970; it encompassed 2- and 4-door sedans, a 4-door hatchback, a 2+2-seat coupe, a roadster and even a minibus. All these versions were to use the same platform and suspension, and be driven by engines with various power outputs and displacements from 1 to 1.6 liters. For reasons of model policy, however, Volkswagen

Varianten des Porsche Typ 1966 (links Roadster, rechts Kleinbus)
Versions of the Porsche Type 1966 (roadster left, minibus right)

Linke Seite: Kunstharzmodell des von Porsche Style gestalteten Typ 1966
Left page: Synthetic-resin model of the Type 1966, a Porsche Style body design

Versuchswagen des Typ 1966 auf der Porsche-Teststrecke in Weissach
Experimental Type 1966 car on the Porsche proving-ground track in Weissach

ben Boden- und Fahrwerksgruppen teilen. Als Antrieb waren Motoren in verschiedenen Leistungsklassen von 1,0 bis 1,6 Liter Hubraum vorgesehen. Aus modellpolitischen Überlegungen entschied sich Volkswagen jedoch gegen den weit fortgeschrittenen Vorschlag.

Die zweifellos bekannteste Kooperation zwischen Porsche und Volkswagen war der im Herbst 1969 vorgestellte VW-Porsche 914. Erste Konzept-Skizzen des Roadsters wurden vier Jahre zuvor von Porsche erarbeitet und 1966 vom Volkswagen-Vorstand als »Entwicklungsauftrag 142« freigegeben. Unter Verwendung möglichst vieler Gleichteile aus der VW-Serienproduktion entwickelte Porsche einen zweisitzigen Mittelmotorsportwagen mit Vier- und Sechszylindermotor, der ab September 1969 von der gemeinsamen »VW-Porsche-Vertriebsgesellschaft mbH« angeboten wurde. Bei der Konzeption des neuen Modells hatten die Stuttgarter Ingenieure eine besondere Schwierigkeit zu meistern. Der 914 sollte sowohl zum VW-Verkaufsprogramm kompatibel wie auch als Porsche erkenntlich sein – ohne dem etablierten 911 Konkurrenz zu machen. Formal nicht unumstritten, zeichnete sich der 914 durch eine sachliche und klare Liniengebung

decided to abandon this project although it had reached an advanced stage.

Probably the best-known example of cooperation between Porsche and Volkswagen is the VW-Porsche 914, introduced in the fall of 1969. Porsche had drawn up concept sketches for this roadster four years previously, and in 1966 it was approved by the Volkswagen Executive Board as “Development Contract 142”. Using as many parts as possible from the standard VW production program, Porsche developed a two-seat mid-engined sports car with a four- or six-cylinder engine. Starting in September 1969, the car was sold by a joint company, “VW-Porsche-Vertriebsgesellschaft mbH”. The new model’s basic concept confronted the engineers in Stuttgart with a particular difficulty: the 914 was to be compatible with the VW sales program, but also identifiable as a Porsche – though without competing directly with the well-established Porsche 911 models. They solved this problem by giving the 914 clearly defined, rather severe styling which was the subject of some controversy. The mid-engine layout was adopted for technical reasons: optimal space

Prospekt der VW-Porsche Vertriebsgesellschaft für den Typ 914
Sales leaflet issued by the VW-Porsche Vertriebsgesellschaft for the Type 914

Ferry Porsche (rechts), Bundeskanzler Kiesinger (links) und VW-Chef Kurt Lotz (mitte) mit dem VW-Porsche 914 auf der Frankfurter IAA 1969
Ferry Porsche (right), Federal German Chancellor Kiesinger (left) and VW Chief Executive Kurt Lotz (middle) with a VW-Porsche 914 at the German Motor Show in Frankfurt (1969)

S-R 2145
D
EBE
RR 7
D

aus. Die Mittelmotoranordnung wurde auch unter technischen Gesichtspunkten ausgewählt: Zum einen bot sie eine optimale Raumausnutzung bei geringer Fahrzeuggröße, zum anderen gewährleistete das Konzept eine günstige Gewichtsverteilung bei tiefem Schwerpunkt. Ein weiterer Vorteil waren die konstruktionsbedingt vorhandenen Knautschzonen vorne und hinten, die in Kombination mit dem integrierten Überrollbügel ein überzeugendes Verkaufsargument waren. Bei einem Grundpreis von weniger als 12.000 DM entwickelte sich der VW-Porsche 914 zu einem Bestseller und war Anfang der 1970er Jahre der meistverkaufte deutsche Sportwagen. Bis zur Produktionseinstellung im Jahr 1976 wurden insgesamt 118.982 Einheiten des VW-Porsche 914 gebaut.

Ursprung des Porsche 924 war der im März 1972 erteilte Volkswagen-Entwicklungsauftrag »EA 425« zum Bau eines VW-Porsche 914-Nachfolgers. Ziel war die Konzeption eines neuen Sportwagens, dessen Herstellungs-, Wartungs- und Ersatzteilkosten durch die Verwendung von VW-Serienaggregaten und -teilen deutlich unterhalb des Porsche 911 lagen. Nach ersten Vorentwürfen fiel die Entscheidung auf die von

Familientreffen:
VW-Porsche 914/4 1.7 und der »VW-Käfer« auf der Autobahn (1970)
Family meeting: VW-Porsche 914/4 1.7 and VW Beetle on a German "autobahn" (1970)

utilization in a relatively small car, excellent weight distribution and a low center of gravity. Another advantage of the design was the ease with which impact-absorbing body zones could be provided at the front and rear; together with the integral rollover hoop, these were a convincing sales argument. At a basic list price below 12,000 Deutschmarks (DM), the VW-Porsche 914 soon became extremely popular, and was the top-selling German sports car in the early 1970s. By the time production ceased in 1976, no fewer than 118,982 VW-Porsche 914 cars had been built.

The Porsche 924 originated from development contract EA 425 issued by Volkswagen in March 1972, for the design of a sports car to succeed the VW-Porsche 914. This was to be a new concept, but one that would use standard VW components and assemblies and therefore cost distinctly less to manufacture, service and repair than the Porsche 911. When the first preliminary drafts had been examined, the choice fell on Porsche's design, which had a front-mounted four-cylinder engine and a transaxle (combined gearbox and final drive) at the rear. An unusual feature of the aerodynamically efficient coupe body was the all-glass tailgate giving

Verkaufsschlager: Der VW-Porsche 914 wurde insgesamt fast 120.000 Mal gebaut

Top-seller: nearly 120,000 VW-Porsche 914 cars were built

Porsche vorgeschlagene Bauweise mit wassergekühltem Vierzylinder-Frontmotor und Transaxle-Getriebe an der Hinterachse. Besonderheit der strömungsgünstigen Coupé-Karosserie war der über eine Ganzglas-Heckklappe zugängliche Kofferraum.
Der »EA 425« befand sich bereits im Stadium der Serienvorbereitung, als die Volkswagen AG das Projekt 1975 aus modellpolitischen Gründen stoppte. Überzeugt von der Qualität der eigenen Entwicklung erwarb Porsche die serienreife Konstruktion und brachte den Sportwagen zum Jahresbeginn 1976 als Porsche 924 auf den Markt. Durch die Verwendung zahlreicher Bauteile aus dem Volkswagen-Konzern konnte der bei Audi in Neckarsulm montierte 924 zu einem Grundpreis von nur 23.240 DM angeboten werden. Aufgrund der weltweit großen Nachfrage stieg die Fertigung bis Juli 1976 auf 80 Fahrzeuge pro Tag an, so dass auf den Typ 924 zum Ende des Geschäftsjahres ganze 48 Prozent des Porsche-Fahrzeugumsatzes entfielen. Bis zum Produktionsende des Porsche 924 S im Jahr 1988 entstanden 150.684 Einheiten des Porsche-Einsteigermodells.

Im neuen Jahrtausend bahnte sich eine weitere Kooperation ihren Weg: Unter der Bezeichnung »Colorado«

VW »EA 425« als Modell in der Porsche-Designabteilung in Weissach (1972)
Model of the VW EA 425 in Porsche's Design Department in Weissach (1972)

access to the trunk. The "EA 425" had reached the series-production preparation stage in 1975, but then Volkswagen AG decided to abandon it for model-policy reasons. Porsche was convinced of the quality of the car it had designed, and decided to go it alone: the design was purchased and the Porsche 924, as it was now known, launched at the beginning of 1976. It was assembled by Audi in Neckarsulm, and the use of a large number of components from the VW parts bin

Porsche 924 aus dem
Modelljahr 1976
*Porsche 924,
model year 1976*

Porsche 924 in der Baustufe 1 auf dem Weissacher Testgelände (1972)
Porsche 924, construction stage 1, at the proving ground in Weissach (1972)

entwickelten Porsche- und Volkswagen-Ingenieure eine gemeinsame Plattform für ein geländegängiges Mehrzweckfahrzeug. Seit 2002 sind der Porsche Cayenne und der VW Touareg sehr erfolgreich im Marktsegment der Sports Utility Vehicles unterwegs. Die enge Zusammenarbeit während der Entwicklung findet auch in der Produktion ihre Fortsetzung. Im VW-Werk Bratislava wird nicht nur der Touareg produziert, sondern auch die lackierten Rohkarosserien des Cayenne werden von dort ins Leipziger Porsche-Werk geliefert. Auch in der Zukunft setzt Porsche auf die Zusammenarbeit mit Volkswagen. Die Rohkarosserien des Panamera, einem viertürigen Gran Turismo, werden im hochmodernen VW-Werk Hannover produziert und von dort zur Endmontage nach Leipzig angeliefert.

permitted a list price of only 23,240 DM. Worldwide demand was strong, so that by July 1976 output had risen to 80 cars a day. At the end of the fiscal year, the Type 924was found to account for a remarkable 48 percent of Porsche's total car turnover. By the time the last Porsche 924 S was built in 1988, 150,684 of these entry-level Porsches had been built.

In the new millennium, further cooperation began to take shape: the code-name "Colorado" was applied to the shared platform developed by Porsche and Volkswagen engineering staff for a multipurpose vehicle with off-road capability. Since 2002, the Porsche Cayenne and the VW Touareg have been highly successful in the Sports Utility Vehicle market segment. This close cooperation during development work was continued when production started: the VW plant in Bratislava not only produces the Touareg but also painted bodies for the Cayenne, which it ships to the Porsche plant in Leipzig. Bodies for the four-door model Panamera are produced at the ultra-modern VW plant in Hanover and shipped to Leipzig for final assembly.

Interieur des Porsche 924
im Vorserienstadium
A pre-production interior
for the Porsche 924

Cayenne Turbo (links) und Cayenne S vor dem Kundenzentrum in Leipzig 2002
Cayenne Turbo (left) and Cayenne S in front of the Customer Center in Leipzig, 2002

Rechte Seite: In der derzeit modernsten Automobilfabrik der Welt in Leipzig montiert Porsche den viertürigen Gran Turismo Panamera gemeinsam mit dem Geländewagen Cayenne auf einer Linie
Right page: In the world's most modern automotive factory to date in Leipzig Porsche is assembling the four-door Gran Turismo Panamera together with the Cayenne sports utility vehicle on one line.

03
E/4
E/4

»Der Käfer ist unser Kind. Ein altes Kind, ohne Zweifel, aber aus vornehmen Hause.«

“The Beetle is our child. It’s getting on in years now, but it comes from a good family!”

DR. ING. H.C. FERDINAND »FERRY« PORSCHE

RODNY DUM /
GEBURTSHAUS
/ BIRTHPLACE
PROF. DR. ING. H.C.
FERDINAND PORSCHE
KONSTRUKTER / KONSTRUKTEUR / DESIGNER
*3.9.1875 MAFFERSDORF +30.1.1951 STUTTGART

Dr. Ing. h. c. F. Porsche AG
Porscheplatz 1
70435 Stuttgart
Germany
www.porsche.com/museum

Inhalt & Abbildungen: Porsche AG, Historisches Archiv
Content & photos: Porsche AG, Historical Archives
Design & Layout: SEIDLDESIGN
Druck / Printing: GZD

Erschienen im DuMont Buchverlag
www.dumont-buchverlag.de

ISBN 978-3-8321-9298-3
Printed in Germany